Migratory Movements of Georgia's Greek Community

LINGUISTIC CONSTRUCTION OF SOCIAL BOUNDARIES: IDENTITIES AND BELONGING

Edited by Konstanze Jungbluth und Mônica Maria G. Savedra

VOLUME 8

PETER LANG

Nika Loladze

Migratory Movements of Georgia's Greek Community

The Impact of Current Socio-economic Transformations

PETER LANG

Bibliographic Information published by the Deutsche Nationalbibliothek
The Deutsche Nationalbibliothek lists this publication in the Deutsche Nationalbibliografie; detailed bibliographic data is available online at http://dnb.d-nb.de.

Zugl.: Frankfurt (Oder), Europa-Univ., Diss., 2019; Datum der Disputation: 5.7.2019; Betreuerin: Prof. Dr. Konstanze Jungbluth, Prof. Dr. Alexander Kartosia

Library of Congress Cataloging-in-Publication Data
A CIP catalog record for this book has been applied for at the Library of Congress.

Research for this book was made possible by the Europa-University Viadrina, Frankfurt (Oder), and a PhD scholarship from VolkswagenStiftung in the context of the research project „The impact of current transformational processes on language and ethnic identity: Urum and Pontic Greeks in Georgia".
The Faculty of Social and Cultural Sciences at the Europa-University Viadrina has generously supported this publication.

D 521
ISSN 2509-4505
ISBN 978-3-631-84726-8 (Print)
E-ISBN 978-3-631-85970-4 (E-PDF)
E-ISBN 978-3-631-85971-1 (EPUB)
DOI 10.3726/b18609

Peter Lang – Berlin · Bern · Bruxelles · New York · Oxford · Warszawa · Wien

This publication has been peer reviewed.

www.peterlang.com

In memory of my Father;
To my mother, brother and my greatest supporters: my nieces Tinatin and Kato, and my nephew Nikoloz

Acknowledgements

Work on this book was accompanied by numerous challenges, but I was lucky enough to be supported by wonderful people on my way to overcoming them.

First, I would like to express my sincere gratitude to my supervisor Konstanze Jungbluth, without whose immense knowledge, patience, encouragement, and continuous support it would have been impossible to accomplish this work. I would also like to express my earnest appreciation to Stavros Skopeteas, whose comments and feedback provided inestimable help in the composition of this study. It is thanks to their guidance that I was able to carry out this research in the framework of the VW-project "The impact of current transformational processes on language and ethnic identity: Urum and Pontic Greeks in Georgia". Furthermore, the involvement and knowledgeable advice of Alexander Kartosia have proven to be crucial to the making of this book.

It goes without saying that I am utterly thankful to all the participants in the project, and especially to Natia Jalabadze, who had already supported me before it began and has played a major role in my professional growth. Together with her, I would like to express my deep and respectful gratitude to Lia Melikishvili, for the pleasure of working with her and for sharing her invaluable experience with me. Ekaterine Kartvelishvili and Violeta Moisidi also played a key role in the successful development of the project by helping me connect with representatives of the small Greek community in Georgia.

I feel very lucky for everything that happened while carrying out fieldwork with Concha Maria Höfler, on this project or otherwise. Our work over the years has developed into a great friendship and an exceptional experience for me. Thank you for all the adventures, arguments, advice and support, and for your hospitality in Berlin, where I always felt at home thanks to you.

A very special gratitude goes out to all the interviewees in Georgia, Greece, and Cyprus, who were kind enough to share their most important personal stories.

As for the English rendition of the book, it would have been impossible without the help of Aleksandre Shervashidze, Mariam Menabde, Dustin Gilbreath and Uri Gordon.

I would also like to thank the Viadrina University and its team: Rita Vallentin, Edgar Baumgärtner, Dorothea Horst, Maria Klessmann, Peter Ulrich and Dominik Gerst. They have always made me feel welcome. I am particularly thankful to Peter Rosenberg for the inspiring conversations we had and to Iris

Franke for the many administrative procedures she dealt with. I take this opportunity to apologize to Frau Franke for not yet having been able to hold a proper exchange in German with her.

I want to thank Tbilisi State University and the Department of Human Geography, where under the guidance of Joseph Salukvadze, Giorgi Gogsadze, and Valerian Melikidze I made my first steps in the academic field and learned to understand current processes "in space and time". I am particularly grateful to Giorgi Gogsadze and Valerian Melikidze for their consultations during the first stages of this work, as well as to Vladimer Chkhaidze, who introduced me to GIS, and David Sichinava for his kind advice and his kindness as a colleague.

Clearly getting through this work required much more than academic support, what I always felt from Anni Chkoidze whom I'd like to thank especially for tolerating my chaotic schedule.

I could not fail to thank the wonderful people I befriended in Germany, who welcomed me with open arms and have made each of my visits unforgettable: Jochen Hänsel, Jakob Kleine, Hugo Lancheros, Jürgen Joseph, Vicky Uhlich, Emil Lamprecht, Taib Hayat and many others whom I have surely omitted. I want to thank especially Sylwia Szmajduch for making comradeship mean such a wondrous thing to me. My thanks for conversations over beer goes to Michael Krengel, a colleague in mapping and a great friend.

I know that none of my Tbilisian friends would mind too much if, instead of a lengthy yet incomplete list, I singled out Giorgi Jibladze and Nodar Davituri, who have heartened me through harsh times and provided me with comfortable working conditions when I most needed them.

Throughout the writing process, I was always feeling the support of Edita Štulcaitė. Same goes to Vakhtang Urushadze and Mako Machavariani for our friendship full of "tannins" and wonderful memories.

Finally, I would like to express my heartfelt gratitude to my family: my mother Lela Mikiashvili, father Ushangi Loladze and brother Luka Loladze, for always being there for me, for their continuous support, and for believing in me; and to my uncle David Mikiashvili and aunt Tamar Lashkarashvili for their significant role in my life.

Contents

List of Maps

List of Figures

List of Tables

Chapter I. Introduction

I first became interested in migration while studying at the faculty of Human Geography at Tbilisi State University. However, my first venture into migration research on the Greeks of Georgia began later. In 2010, Natia Jalabadze suggested that I participate in the Volkswagen Foundation-funded research project, "The Revitalisation of Traditional Law in Georgia", where I served as a fieldwork assistant to Stéphane Voell in the Kvemo Kartli region.

The project aimed to study traditional law practices among Svan eco-migrants, who were relocated from western Georgia's highland region of Svaneti to a region compactly populated by Greeks (Voell, Jalabadze, Janiashvili & Kamm 2014; Voell 2016). The project allowed me to become closely acquainted with the region's multiethnic background, and in particular with the changes and ethnopolitical processes that occurred after Greek emigration. Most importantly, I had the opportunity to talk for the first time with the *Pontic* and *Urum* Greeks living in Kvemo Kartli, about whom I possessed very scarce information beforehand.

The chief difference between these two Greek communities living in Georgia is linguistic: the Pontic Greeks are a graecophone group and speak the Pontic dialect of the Greek language, while the turkophone speakers of an eastern Anatolian Turkish variety are labeled as (Caucasian) UrumGreeks. Both groups consider themselves Greek and belong to the Eastern Orthodox Church, which is of particular importance for their identity construction (Höfler 2020, Sideri 2006).

This study, which focuses on the emigration of Georgian Greeks, was made possible through my participation in the Volkswagen Foundation-funded project, "The impact of current transformational processes on language and ethnic identity: Urum and Pontic Greeks in Georgia", led by Konstanze Jungbluth (European University Viadrina) and Stavros Skopeteas (University of Bielefeld) (see Höfler 2020; Höfler, Böhm, Jungbluth and Skopeteas 2016). The mass emigration of the Greek population from Georgia in the 1990s, immediately after the collapse of the Soviet Union, is linked to the momentous transitional processes of the same period. The country met independence unprepared and found it hard to deal with both internal and external challenges. Concomitantly with increased nationalism, civil war,[1] and the ongoing military conflicts surrounding territorial

1 This refers to the period of armed confrontations between supporters of Georgia's first president Zviad Gamsakhurdia and the United Opposition in 1991–93.

integrity, [2] the economy also suffered a sharp decline, unemployment rose, and inflation reached record levels. After the fall of the Iron Curtain, labor emigration occurred *en masse*, with direct impact on the ethnic composition of the country. Although the population outflow was all-encompassing, census data show that Georgia's Greek population was the most active in the emigration processes, with their share in the country's population falling from 1.9% in 1989 to 0.1% in 2014. In absolute terms, the Greek population declined by 95%, from 100,324 in 1989 to 5,500 in 2014.

This work is dedicated to the study of the interaction between these complex transformational socioeconomic processes, migratory vectors, and the role of ethnic belonging, with the goal of outlining the under-examined determining factors driving the mass emigration of the Pontic and Urum Greeks of Georgia.

The rest of this chapter briefly presents my research questions (1.1), the current state of research (1.2), methodology and sampling (1.3), and the book outline (1.4).

1.1 Research questions

In the study of the migratory processes that occurred in the post-Soviet countries immediately after independence, focus is given to co-ethnicity, regardless of the specific challenges facing particular countries. During this period, the return of ethnic groups – scattered across the former Soviet empire – to their "ethnonational homeland", either by choice or by force, is considered to be one of the dominant factors driving migration processes in the post-Soviet space (Brubaker 1998; Ivakhnyuk 2006).

Skeptical about interpreting these processes from a purely ethnonational perspective, I set out to study the ways in which economic factors influenced the decision to emigrate. As a result, I chose to analyze the process of the mass outflow of Greeks from Georgia in the framework of international labor migration. To do so, I utilize labor migration theories and interviews conducted during fieldwork.

To achieve this goal, it was necessary, on the one hand, to delineate how recent socioeconomic processes stimulated Georgian Greeks' labor emigration, and on the other hand, to research the importance of emigration routes and the role of the ethnic factor or "return/repatriation" to Greece.

2 Military conflicts which started in Georgia's autonomous regions of South Ossetia and Abkhazia in 1990–93.

In light of the above, the research hypothesis has been conceived as follows: in the context of the harsh economic situation that arose in post-Soviet Georgia, the emigration of Greeks must have been primarily determined by economic factors, and the prospect of finding employment abroad was the main motivator for emigration, as was the case with other Georgian citizens. Ethnic affiliation also influenced the decision-making process, albeit not as a source of desire for repatriation or return to the "homeland", but rather as a supporting mechanism for emigration, considering Greece's ethnic immigration policies.

The above hypothesis has been examined in Georgia based on research conducted in the regions with compactly populated Greek settlements, and is not centered on one specific labor emigration theory. Instead, the book applies a theoretical synthesis, discussed in the next chapter.

1.2 Current state of research

Although large-scale emigration from Georgia began as soon as the country gained its independence – leading to a corresponding increase in the number of studies of the topic – to this day there is scarce academic literature about the Greek community of Georgia and its migration processes.

From an anthropological point of view, I use the work of Pashaeva (1983; 1992), Sideri (2006; 2007; 2012), Jalabadze (2011) and Melikishvili's (2016). They mainly deal with the Urum population of eastern Georgia. The book "Pontic Greeks in Adjara – Past and Present" (Giorgadze et al. 2017) is an important source of information about Pontic Greeks from western Georgia. Additionally, the study draws on the dissertation of my co-researcher, Concha Maria Höfler (2020). There is a larger number of sources on the history of Greek immigration to Georgia, and I compared the works of Georgian (Kuakhchishvili 1942; Jaoshvili 1984; 1996; Garakanidze 2000) and Greek (Kitromilidis 1990; Hionidu 2012) researchers.

Academic research on the period of Soviet repression and deportations is particularly scarce, not only concerning the Greeks of Georgia but more generally. In this regard, it may represent the least understood period of recent Georgian history. While discussing deportations (Chapter 4), I mainly use research by Junge, Bonwetsch and Muller (2015) and Junge, Tushishvili, and Bonwetsch (2015). For the total numbers related to these processes, I rely on sources from prominent Sovietologists (Courtois et al. 1999; Phol 2000; Rummel 2017). Popov (2016) carried out a study specifically on North-Caucasian Greeks.

The academic literature on current labor migration in Georgia focuses less on issues related to ethnic minorities and diasporas, and mainly revolves around statistical data (Gachechiladze 1999; Sulaberidze 2017), the impact of emigration on the country and regions (Chelidze 2006; Zurabishvili 2007; Shelia 2016), money transfers (Archvadze 2017), and gender (Lobjanidze 2009; Iashvili et al. 2014; Shelia 2017).

The theoretical models and literature concerning international labor migration are discussed in detail in chapter II.

The main contribution of this study is a theoretical and substantive contribution to our understanding of the under-examined migratory processes of Georgian Greeks, which in turn may be extended to the Greek Diaspora in the post-Soviet space more broadly. In addition, since the research analyzes the empirical and theoretical dimensions of labor emigration, it makes a significant contribution to the study of labor emigration from Georgia as a whole.

1.3 Methodology and sampling

In planning and implementing the research project, I faced two substantial challenges. The first was the lack of statistical data, as official statistics on external migration processes in Georgia lack reliability, and researchers working on issues of migration and demography have to use their own or other experts' calculations (Vadachkoria 2004, Zurabishvili 2007).

For example, according to the census data of 1989 and 2002, the population of Georgia decreased from 5.4 million to 4.371 million. According to the same data, only 113,000 people moved abroad, a difference which obviously cannot be explained by mortality alone (Zurabishvili 2007: 9). [3]In the case of Greeks, while census data obviously indicate mass emigration, with numbers decreasing from 100,324 in 1989 to 5,500 in 2014, I did not have statistical information about their emigration routes.

In addition, the dramatic decline in the number of Greeks in the country posed other important challenges. First, to conduct fieldwork, I had to find members of the small Greek community who remained in Georgia. Second, I had to choose respondents in such a way that the data obtained would be useful for a quantitative approach.

Consequently, I had to make alterations to the methodology that was initially planned, which was mainly based on my quantitative questionnaire. After some adjustments, I decided that using a case study as a methodological basis for the research would be the optimal choice, especially since during the fieldwork,

---

3 Demography and statistical materials are discussed in Chapter III.

together with Concha Höfler, we also collected data using her qualitative questionnaire.[4]This change had an entirely positive effect on the study, because the data obtained through the qualitative questionnaire allows for an in-depth analysis, while the information obtained through the quantitative questionnaire, despite the small number of respondents, reveals interesting trends, which can be used to verify, enrich, and compare data obtained through the qualitative data collection and analysis.[5]Case studies have been used in the social sciences since the 19th century. It is believed that Frédéric Le Play, the French founder of social economics, used this method for the first time in 1829 to study family budget management (Forni, Fray & Quaranta 1998). Since the late 1960s, after Barney Glaser and Anselm Strauss (1967) developed Grounded Theory and its methods, case studies and qualitative methods more generally have attracted an increasing number of scholars and are no longer overshadowed by quantitative methods (Tellis 1997).

Case studies make it possible to thoroughly study a particular case, community, or group by analyzing its interrelationships. The main data collection techniques are participant observation, in-depth interviews, focus group discussions, and/or video recording. When conducting a case study, it is vital that the researcher spends extensive time in the field to become better acquainted with the life of the researched group, to share its members' everyday lives, and to gain their trust.

For my research, one of the advantages of this method is its flexibility, as its final design is developed and refined during the actual research process, according to the information obtained. Moreover, the researcher has the opportunity to triangulate both qualitative and quantitative data. Through such a combination, on the one hand, the research data is further enriched, and on the other hand, the researcher is able to analyze it from different angles and reexamine the results.

1.3.1 Semi-structured questionnaire

As noted above, the study uses both qualitative and quantitative data. Qualitative data was collected using a semi-structured questionnaire, developed by Concha

4

5 I became convinced of the appropriateness of this approach especially after reading Tamar Zurabishvili's dissertation, which uses a case study perspective and describes the determining factors for labor emigration using the example of the East Georgian town Tianeti very thoroughly (see Zurabishvili 2007)

Maria Höfler (Höfler 2020: 81). It consists of 11 thematic blocks and emphasizes the following issues:

- Narratives of how "the Greeks" first came to Georgia, how the interviewees' grand-parents had lived in their youth, how life was during the Soviet Union, the changes in the years since the end of the Soviet Union, and Georgia's independence;
- Whether there had been any discrimination on ethnic grounds during the Soviet Union or after;
- Explanations for the mass Greek emigration out of Georgia and personal and family experiences thereof;
- The (conflict prone) internal migration to Ts'alk'a in the early 90s and the situation there today;
- Language competence, use, and evaluations of the [interviewees], in their families, their community, and "society";
- The interviewee's sense of belonging and perception of inter- and intra-communal boundaries;
- The interviewee's and the community's religious and cultural practices; and
- The various Greek organizations operating in Georgia today and the interviewee's relationship to them.

Based on the questionnaire, the respondents had the opportunity to answer with their life stories and to share their experiences. This helped collect exhaustive data. This approach was particularly effective in describing the decision-making process behind emigration, because, as it turned out and as discussed in Chapter V, the influence of complex factors emerged, whose interconnection would not have been found through a quantitative questionnaire alone.

The interviews did not consistently preserve the order of the aforementioned thematic blocs. This enabled both us interviewers and the interviewees to introduce and discuss new topics during conversations. This allowed topics to emerge, which are relevant to the research, but for various reasons were not considered during the planning stage. The best example was when during the actual fieldwork, some of the respondents spoke extensively about the Soviet deportations, as discussed in chapter IV.

1.3.2 Quantitative questionnaire

The main purpose of the quantitative questionnaire was to collect socio-economic data about Greeks living in Georgia. Questions were related to living conditions, employment, income sources, educational background, migration background, factors driving migration, local and transnational social networks, and the impact of locale and destination countries on the migration process.

The questions' response options took the following forms:

- Yes/No questions;
- Scaled questions;
- Multiple choice questions;
- Open-ended questions.

Since it was impossible to obtain similar information about Greeks who emigrated from Georgia using official data and/or other studies, this questionnaire provided the opportunity to interview emigrants' family members and other relatives who stayed in Georgia (substitute respondents) about their family members who had emigrated. Nonetheless, this approach has its own shortcomings. For example, some of the relatives we interviewed did not have information or did not want to answer every question on the survey about family members that emigrated. In such cases, these components of the interviews were considered invalid and dropped from the data. Still, I was able to obtain valuable information about the socio-economic situation of the families who had stayed in Georgia.

1.3.3 Sampling and data processing

Within the framework of the research, most of the fieldwork took place in Georgia, in the regions formerly or still compactly populated by Greeks. These include rural and peri-urban areas in Kvemo Kartli, Samtskhe-Javakheti, and Adjara as well as in the large cities Tbilisi and Batumi (see Map 1).[6] The first stage of the fieldwork took place in Spring 2013. It lasted four months, and included the regions Ts'alk'a, Tetrits'q'aro, Borjomi, Kobuleti, Batumi, and the capital Tbilisi. The second stage lasted for two months in Spring 2014 and took place in Tbilisi, Ts'alk'a, Kobuleti, and Batumi. Additionally, we visited Greece and Cyprus for one month in June 2014. Here, I would like to mention that I conducted every stage of the fieldwork together with Concha Maria Höfler, which both made the work easier and the process of obtaining data especially productive.

Throughout the fieldwork period, we familiarized ourselves with the living conditions in these regions, met with local municipality employees, and recorded

6 Fieldwork could not be carried out in Abkhazia, occupied by the Russian Federation as a result of 1992–93 and 2008 wars and thus off-limits for Georgian citizens. In any event, the absolute majority of Greeks had already left Abkhazia during and after the 1992–3.

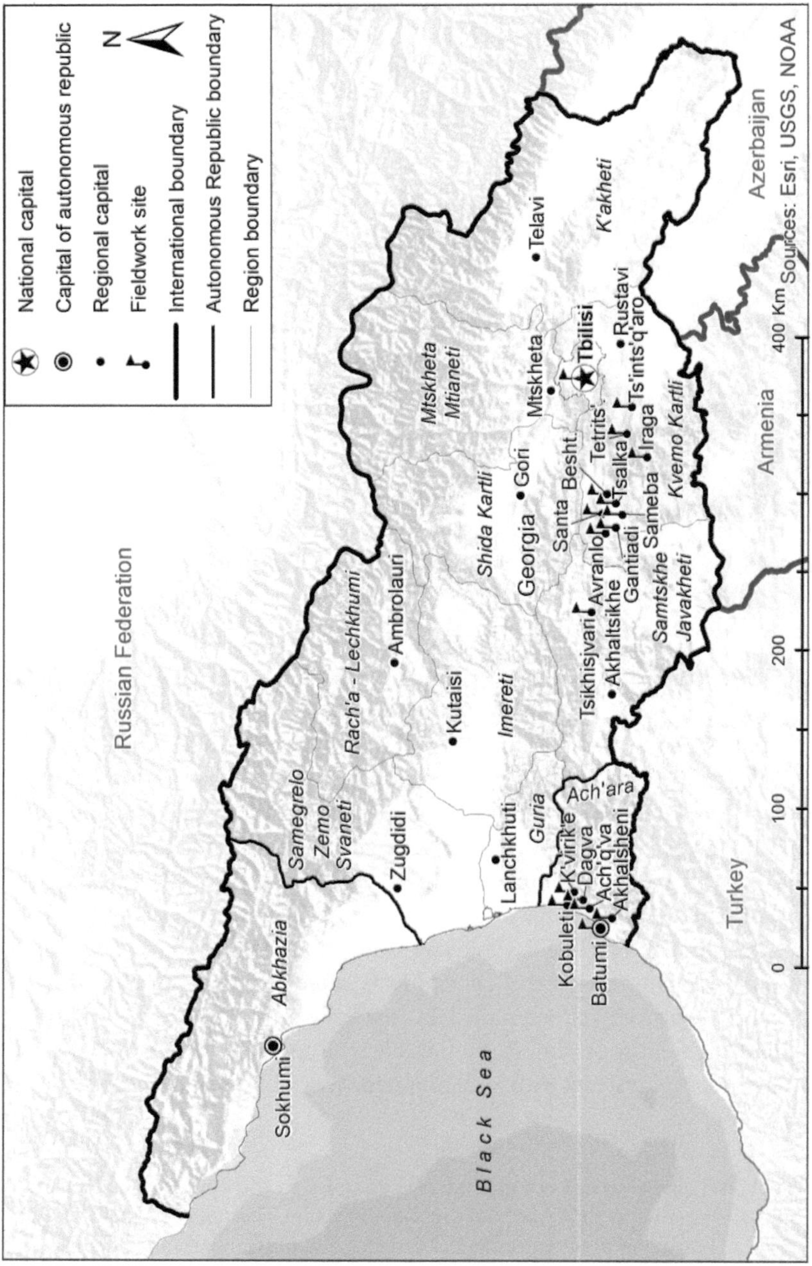

Map 1 Fieldwork sites

interviews with both Greeks and Georgians (Svans, Adjarians). After a preliminary analysis, I chose to include a total of 47 interviews in my analysis, including 22 Urum (10 male, 12 female) and 25 Pontic (12 male, 13 female) Greeks. Their ages ranged from 19 to 77 (average 45.7) in the case of Urum respondents and 19 to 81 (average 49.3) in the case of Pontic respondents.

The interviews were carried out in Russian and Georgian. For the analysis in Chapter V, I have chosen to present an idiomatic English translation of the interviews for the sake of clarity. While presenting interview excerpts within this book, I chose to assign random acronyms to interviewees, thus placing them on equal footing with Concha Höfler (CH) and myself (NL).

The spatial distribution of the interviews is as follows. Regarding Urum respondents, we recorded 13 interviews in the Ts'alk'a municipality and nine in Tbilisi. We talked to most of the Pontic respondents in Adjara: nine in Batumi and six in Kobuleti municipality. In eastern Georgia three interviewees were recorded in Tsikhisjvari, five in Tetrits'q'aro municipality, one in Ts'alk'a, and one in Tbilisi.

Because the study was related to a quantitatively small ethnic minority group, we used the snowball method for sampling, since there was no adequate list which could be used as a sampling frame. As a result, finding respondents became easier over time. In cities, we mostly used personal contacts, after which we would contact other local Greeks with references from the interviewee. In the countryside, we asked about local Greeks in village shops and convenience stores. We also visited local administration offices, a strategy which turned out to be largely successful. During these visits, we introduced ourselves, told local government officials about our work, and collected information about their districts. This approach was very successful in Ts'alk'a, where the head of the local administration provided us with information about the ethnic composition of the settlements and contact information for local Greeks.

The local population also assisted us in finding respondents based on their personal contacts, and always tried to refer us to "more competent individuals" who were knowledgeable about the Greek community and the history of the region. It is worth noting that local taxi drivers were exceptionally helpful in this regard. They knew the region well and introduced us to many respondents through their personal connections.

Despite the fact that we had two relatively extensive questionnaires, the interview process went well. The respondents' level of trust was high, and on many occasions, they tried to provide details and additional background information about certain events to Concha Höfler, since they viewed her as an outsider. Complementing this was the fact that I was Georgian and a native speaker, which often helped us overcome cultural barriers. Each interview was conducted with both of the abovementioned questionnaires, and lasted between 30 and 90 minutes. The participants responded more easily to the quantitative questionnaire after answering Concha Höfler's semi-structured qualitative questionnaire, which had already helped to build trust.

The most difficult part of the interview process was maintaining an age balance. Because of the high migration rates, it was extremely difficult to find Greek interviewees between the ages of 18–30. People in this age group are most active in seeking employment, and the majority of them work abroad. In contrast, there was a high proportion of people over 60 who stayed behind in Georgia and took care of their households. In some cases, factors preventing the migration of these elderly people include their attachment to their home, their health conditions, linguistic barriers, and personal circumstances. Some of them prefer to live in Georgia due to the lower cost of living.

Besides Georgia, our visit to Greece and Cyprus was extremely beneficial. Even though we only had one month, and it was therefore impossible to conduct substantial research, we gained a comprehensive picture of local standards, the living conditions of emigrated Georgian Greeks, and their role in Greek societies. Our visit to the Historical Archive of Refugees in Thessaloniki was also very useful. Meetings with Dr. Eleni Sideri in Thessaloniki, and with Professor Niyazi Kizilyurek, Dr. Christiane Bulut, and Dr. Dionysius Zoumpalidis in Nicosia, Cyprus, who shared their experiences and views on migration and post migration processes, were of great importance.

Finally, the study includes 25 additional interviews obtained by another project member, Ekaterine Kartvelishvili, using our quantitative questionnaire, in March 2015 in Greece. Since in Greece, similarly to Georgia, it is impossible to carry out a comprehensive quantitative analysis using such a small amount of data, these interviews are used only to highlight trends and enhance the present analysis. The interviews obtained in Greece are only presented in subsection 5.1 and on maps 10 and 11. They do not pretend to be scientifically accurate in every respect.

Alongside the research work, the information obtained during the fieldwork was incorporated into a unified MS Excel database, into which I later transferred all the processed data from the quantitative and qualitative questionnaires.

The transcription of the qualitative data was conducted using the *Partitur Editor* of the software package *EXMARaLDA*. In addition to being open-source and offering a user-friendly search function when working on a text, this software also allows data to be exported in different formats, which makes it compatible with other software for further analysis (for further information see Höfler 2017; 84–90).

After examining quantitative and qualitative data, I used geographic information systems (GIS) tools to create a visualization of the emigration and its accompanying processes, further supplemented with cartographic methods. Using ESRI software products (ArcMap & ArcCatalog 10.2.2) I compiled geographic data, analyzed information, and built and managed geographic information in a database. This allowed me to spatially perceive and analyze specific routes of emigration, internal/external group connections, and their determining factors.

1.4 Outline

The book is structured as follows. The second chapter presents the theoretical approaches considered to be the most important strains in contemporary research on labor migration. It discusses each theory, and its application to Georgia and the emigration processes of Georgian Greeks.

The third chapter discusses the necessary historical background, describing the spatial characteristics of Georgian Greeks' compact resettlements as well as socio-cultural and demographic processes within them. The combination of these provides a detailed picture for further analysis.

The fourth chapter describes the historical background of the repression and mass deportations that took place under the Soviet Union. The deportations directly affected the Greek population living in Georgia, and significantly changed the ethnic composition of the regions which became compactly populated by Greeks as a result of forced migration.

The fifth chapter presents the analysis of materials obtained during the fieldwork and explores the mass emigration processes that began in the 1990s. The causes of this emigration, emigration vectors, and the factors impacting them are analyzed in the theoretical context of labor emigration.

Finally, in the sixth and seventh chapters, a discussion and conclusion summarize the findings of the study.

Chapter II. Theoretical framework

Due to its multifaceted nature, the study of international migration requires different perspectives of analysis. Thus, it should be considered a subject for interdisciplinary research. In addition to human geography, migration is a subject of research in the fields of sociology, political science, the humanities, and the natural sciences. This is a consequence of migration's significant influence on international and local political processes, demographics, the economy, and culture. Due to this complexity, as of today there is no single approach or theory that fully covers the study of migration and its related processes (Zurabishvili, 2007).

While describing the heterogeneity of migration, Portes underlines four main research subjects: the origins of migration, the direction and continuation of migration flows, the use of immigrant labor, and migrants' socio-cultural adaptation (Portes 1999). To study these subjects individually and exhaustively, different approaches and levels of analysis are required. In this regard the "middle-range" theories, which focus on one or two subjects, are the most effective, since a universal approach that would comprehensively cover all four subjects has not yet been developed (Arango 2000, Massey 1999).

There are many theoretical models for the study of the factors forcing people to migrate. These theories contain a variety of concepts, assumptions and levels of analysis, and operate on different scopes. Most of these theories were created on the basis of specific empirical observations, and developed in relative isolation within their respective disciplinary silos (Arango 2000; De Haas, Castles and Miller 2019). Despite the abovementioned differences, these approaches are not contradictory or incompatible. On the contrary, a theoretical synthesis provides a more complete picture, where various approaches and hypotheses become complementary.

In this work, I discuss five international theories of migration, including neoclassical economic theory (macro and micro), the new economics of labor migration, dual/segmented labor market theory, world systems theory, and migration networks theory.

2.1 Neoclassical economic theory

To this day, neoclassical economic theory remains the most widespread approach. It originated from modeling internal labor migration processes, which occur during phases for economic development (Lewis 1954; Ranis and Fei

1961). This theory examines migration at the micro and macro levels, primarily focusing on migration's economic characteristics. According to neoclassical theory, international migration, like internal migration, is largely caused by geographic and spatial differences in terms of labor supply and demand. The theory also pays significant attention to factors that *push* and *pull* potential emigrants to leave their home countries (Todaro and Maruszko 1987).

Russel King (King 2012) views push and pull factors, and the neoclassical theory as a whole, as a continuation of the British cartographer Ravenstain's approach in "The Laws of Migration" (Ravenstain 1885, 1889). According to King, while studying migration, attention should be paid to this work, which despite being outmoded and the criticism it has faced, is also something of a cornerstone upon which geographers of migration (among others) build new theories. For instance, according to Samers (Samers 2010), Ravenstain's writings should be regarded not as actual laws, but rather as empirical generalizations based on British or other census materials, which cover internal rather than international migration. Furthermore, according to White, Woods, Boyle, and other geographers, these "laws" have laid the foundation for a geographical approach to migration studies, which consequently aided the development of new hypotheses and the further theorization of the subject (White and Woods 1980, Boyle et al. 1998). "The Laws of Migration", based on King's paraphrasis, it can be divided into seven laws (King 2012):

1. The majority of migrants move over short distances; those who move longer distances tend to choose large industrial and commercial hubs as their destination.
2. Migration usually flows from villages to cities.
3. Large cities grow at the cost of immigration, rather than by natural population growth.
4. Industry, commerce, and transportation have a large impact on migration movements.
5. Every migration flow generates a return flow.
6. Women tend to be more involved in short-distance/internal migration, while men constitute the majority of international migrants.
7. Economic reasons are the main cause of migration.

Since Ravenstain's laws have played a major role in the study of migration, discussing them allows us to see historical aspects of how migration is theorized. For example, the first law can be posited as suggesting a gravitational model of migration, derived from Newton's law of universal gravitation: the attraction between two particles (in this case, places) is directly proportional to their masses (populations) and inversely proportional to the square of the distance between them (White and Woods 1980: 39). The second and third laws echo the most widespread migration form in most countries: migration from rural areas to urban ones. This remains the primary direction to this day. In the fourth law, migration is linked to general development and is almost a century ahead of the "Hypothesis of the Mobility Transition", also known as the "Model of Migration Transition" (Zelinsky 1971). The fifth law is the basis for the study of dual migration dynamics, with the use of concepts such as migration balance and return migration. Notably, detailed study of return migration only began in the 1970–80s and is considered understudied to this day. As for the sixth law, it is doubtlessly innovative, since the gender aspect of migration was neglected for almost a century. Finally, the seventh law emphasizes the well-known economic factor, representing the most common form of migration (King 2012).

The push and pull model can be considered a successor to Ravenstain's laws and an attempt to link individual rational choice and structural, rural-urban, and development inequalities. The simplified model views the reasons behind migration, whether on a national or a regional scale, as a set of push factors – poverty, unemployment, shortage of land, political repression, low social status, etc. Consequently, in the destination country, the pull factors appear as the opposite, and/or as otherwise attractive – high wages and employment opportunities, better education, healthcare, political freedom, etc. Everett Lee's version of this model (Lee 1966) utilizes the notion of an "intervening obstacle" between push and pull factors. These obstacles, which the emigrant has to overcome, might include distance, travel expenses, cultural barriers (language or a different lifestyle, political obstacles), international border crossing, immigration regulations, etc. In Lee's theory, besides the intervening obstacles, personal factors are also of importance and indicate that different people, based on personality, economic situation or stage of life, react differently to push and pull factors. For example, for a single, economically active potential emigrant, employment and wage conditions are essential, and the state of the education system in the receiving country is less important. In contrast, education would matter much more to a family planning to emigrate from a country (Lee 1966, King 2012).

From the 1950s to the end of the 1970s, this approach had an enormous impact on the study and theorization of migration, and was adopted in the neoclassical economic paradigm, which is based on profit maximization, rational choice, and workforce mobility (King 2012). The neoclassical model examines migration on the *macro* and *micro* levels.

At the macro level, the cause of emigration is the uneven distribution of labor in space, which along with other factors is in direct correlation to production and especially industrial capital. For example, in countries with an excess of workforce and lack of capital, labor compensation is low, unlike countries with more capital, but a lacking workforce. As a result, workforce starts to flow from low-paid countries to high-paid countries, which changes the dynamics of demand and supply between them. In this regard, migration appears as a positive phenomenon. On the one hand, it reduces the excess workforce in the homeland, and on the other, it eliminates the lack of workforce in the receiving country (Massay et al. 1998).

The microeconomic approach explains migration as an individual choice, in which the potential emigrant, after calculating the costs and benefits, decides to leave their homeland in hope of financial gain. With this approach, international migration becomes an investment in human capital (Sjaastad 1962). People who decide to emigrate choose countries where they can use their labor skills most productively and for better wages. In cases of illegal emigration, migrants choose countries where deportation is less likely to occur. However, before reaching the destination country, migrants are forced to take on material and immaterial expenses: the cost of travel, living costs before finding employment, as well as those associated with learning a foreign language and culture, adapting to the new labor market, severing old bonds and establishing new ones. Consequently, the potential emigrant, after all these calculations, chooses a country where the probability of gaining profit is high (Massey 1999, Todaro and Mazurko 1987).

As previously mentioned, no single theory, including the neoclassical one, can be regarded as a self-sufficient approach for comprehensively explaining the factors contributing to migration. Today, despite criticism, neoclassical theory remains the most widely embraced. Its weakness is considered to lie in the unsuccessful attempt to explain, on the one hand, why despite the push factors around them, there are so few people involved in emigration, and on the other hand, differences in migration balance between countries that are at the same respective stages of structural and economic development. This is due to the scope of the approach, which mostly focuses on micro and macro-economic characteristics, and less so on personal or general socio-cultural factors, the political realities of a given country, or its colonial past, due to which specific countries and regions are characterized by different migration behaviors and directions. These shortcomings require us to find and incorporate other models and approaches (King 2012, Arango 2017, Massey 1998).

2.2 The new economics of labor migration

The new economics of labor migration (*new economics theory*) has played a significant role in the theorization of migration since the 1980s. According to Russel King, it is important to emphasize the significance of Oded Stark's work on the formation of the theory itself (King 2012; Stark 1991; Lucas and Stark 1985; Stark and Bloom 1985).

As opposed to neoclassical economic theory, new economics theory does not consider migration a mere individual choice but presents a model of workforce diversification by a family and/or household to minimize expected or existing economic threats (Massey et al. 1993). Accordingly, new economics theory can be divided into two parts. On the one hand, the decision to emigrate, or who goes where and for how long, is not an individual decision but is agreed upon with the family and/or household members. Sometimes this move from the macro level to the meso level and "the extended family" becomes involved in the decision-making process. In some cases, even the local community is engaged (Massey et al. 1999). On the other hand, the rational choice of emigration is based not only on profit maximization prospects, but also on the distribution of labor and diversification of income. The risk minimization factor is similarly important, especially in poor economies where, during "market collapses", insurance, loans, or savings are not available or are insufficient to compensate for losses (for example due to a bad harvest, natural disasters, economic crisis or high unemployment rates).

There are also differences between neoclassical and new economics theory in their approach to return migration. Neoclassical theory largely disregards return migration. The theory considers it to be an individual's failure, stemming from an incorrect calculation of the costs required for migration and expected benefits. However, according to new economics theory, return migration is perceived as a successful calculation, meaning that the emigrant has reached their goal in the destination country and therefore returns home to invest financial gain/savings for the family's benefit (Cassarino 2004, De Haas et al. 2015).

Not unlike other theories, new economics theory is subject to criticism, as it is effective chiefly in relation to poor, rural households. Moreover, the decision-making process is presented against a backdrop of a harmonious family relationship, where disagreement is out of the question, and the rigid patriarchal culture's influence or inner family conflicts, which undoubtedly affect decision making, are not taken into account. Most importantly, new economics theory does not explain emigrations of the entire family, focusing only on examples of the emigration of one or several members (Arango 2017). Nonetheless, it should be noted that for the study of labor emigration processes in Georgia, new economics theory appears to be a suitable approach and offers relatively appropriate explanations (Zurabishvili 2007).

2.3 Dual/Segmented labor market theory

In this subsection, I will touch upon dual/segmented labor market theory, which together with world-systems theory (see 1.4) is a historical-structural model based on the Marxist interpretation of capitalism. Dual/segmented labor market theory connects migration to structural changes in the economy. However, in explaining the dynamics of migration, this theory focuses on the labor market's demand, and its impact on migration movements (Massey et al. 1993, King 2012).

Segmented labor market theory was connected to migration processes in the 1980s, its main founder being Michael Joseph Piore (Piore 1979). According to Piore, since international migration processes are determined by labor market demand, the main factors driving international labor migration are pull factors as opposed to push factors. Thus, the non-prestigious, low-paid jobs generated in the economies of developed countries, which local residents refuse to accept because of their negative social status implications, emerge as pull factors for potential immigrants from poor countries.

Therefore, in economically-strong receiving countries, the labor market, based on social class, is divided into two primary components and acquires a dual, concurrently segmented character. On the one hand, the market of the receiving country has high-paid, prestigious jobs, which are mostly occupied by local workers; on the other hand, low-paid, non-prestigious positions are held by immigrant workers. As a result, the receiving country always has demand for cheap immigrant workforce, supplied by residents of countries with an underdeveloped or developing economy. The existence of this phenomenon in labor markets is also demonstrated by the fact that under conditions of structural inflation, employers seek to attract low-paid, immigrant workers instead of increasing salaries for locals (Massey et al. 1993, Zurabishvili 2007).

2.4 World-systems theory

World-systems theory likewise belongs to the historical-structural model and offers a more complete and complex approach to explaining migration. According to this theory, the global economic system plays a role in driving migration. According to Wallerstein (1974, 1979), world-systems theory analyses migration in the framework of global capitalism, which began to proliferate in the 16th century, reaching its peak in the 1900s. Later, in the era of decolonization after the Second World War, it acquired a neocolonial and corporate capitalist nature. It should be noted that colonialism had a significant impact on the flow of international migration, which was determined by the established cultural, linguistic, transportation, infrastructural, or administrative links between the former metropole and colonies (Morawska 2007:3). Wallerstein (1974) offers a description of these links by dividing the countries based on their positioning in the global market economy. According to this classification, the dominant capitalist group includes the countries of North America and Europe, Japan, Australia, and New Zealand, which are considered "core" countries, while countries with poor economies are considered "periphery" countries and are dependent on the "core", with asymmetric market ties, capital inflow, and migration. In addition to "core" and "periphery" countries, Wallerstein also classifies "semi-periphery countries" according to their economic development and degree of dependence on "core" countries.

As we can see, according to world-systems theory migration is a result of capitalist development. It stems first and foremost from the colonial experience, when entrepreneurs from "core" countries opened factories in "periphery" countries to take advantage of cheap labor, raw materials and other resources, thus maximizing profit. This economic activity became the basis for strong relations among countries, which is why migration flows are often directed from the "periphery" to the "core". This is how emigration routes were formed, for example, from India and Pakistan to Great Britain, from the Philippines to the United States, and from Algeria to France (Zurabishvili 2007).

It should be noted that, as a rule, traditional societies are less likely to become involved in international migration processes. Their involvement increases, however, with the increase in their participation in global processes and political or trade relations. An instructive example is Georgia's post-independence increase in emigration.

2.5 Migration networks theory

A migration network is a collection of ties between people, which connects migrants, former migrants, and non-migrants, in their home or receiving countries, with the help of their relatives and friends, and is based on their common origin (Bourdieu & Wacquant 1992). These links can be understood as a form of *social capital*, which spreads in migration spaces. Thanks to accumulated information, these networks simplify the international migration process by reducing the related expenses and risks (Massey 1999).

The study of migration networks as a new phenomenon started in the beginning of the 20th century. In this regard, the work "The Polish Peasant in Europe and America" (Thomas and Znaniecki 1918–1920) is noteworthy. This work became the basis for the study of chain migration, the best example of which is migration from rural areas of Italy to American cities' "Little Italies" (MacDonald & MacDonald 1964). More recently, migrants' social networks have come to feature in studies of "transnational" phenomena (Brettel 2008:125).

Social capital, upon which the migration network is dependent, is a sum of material as well as virtual resources that an individual or a group possesses, including partly institutionalized relationships based on mutual familiarity and recognition (Bourdieu & Wacquant 1992). One of the major and noteworthy features of social capital is that it can easily be converted into other forms of capital, particularly financial capital (Zurabishvili 2007).

As noted above, migration networks, with the help of social capital, help reduce the costs and risks of emigration. For example, for initial emigrants who do not possess social ties in the receiving country, the costs of emigration are much higher, especially with illegal migration. Later, when the first migrants establish roots in the receiving country, the costs for potential migrants who are family members, relatives, or other kin is greatly reduced. Each new migrant creates their own micro-network in the receiving country, based on ties to non-migrants and migrants. These relationships later transform into kin or friendship relations that imply responsibilities and support. Furthermore, they create easier ways of acquiring employment and establishing oneself in the receiving country.

Well-developed migration networks make employment in the receiving country's labor market more accessible to network members. Thus, the sustainable development of these networks, which is connected to a progressive reduction of costs, can also be theoretically linked to a progressive reduction of risks, since membership in migration networks allows for sustainable financial, material, and/or immaterial guarantees, and reduces the risk of failure in the receiving country (Massey et al. 1993, Massey 1999).

While migration networks theory and the social capital model do not explain the initial reasons for migration, they are important for explaining the continuation of the migration process and provide insight into the factors and characteristics that influence migration growth.

2.6 Theoretical synthesis

The study of international labor migration theories makes it clear that a combined approach is necessary for researching migration issues, as each theory provides insights into different pieces of the migration puzzle. Moreover, they provide less than satisfying explanations of cause-and-effect connections.

However, the various approaches these theories take do not explicitly contradict each other. It is entirely possible for an individual to become involved in migration as a result of calculating costs and benefits. Risk reduction or the need to pay back loans might drive migration by members of a household. Social connections might have an impact on the individual and family levels. The state's internal or external economic processes might create favorable conditions for migration. Even migration itself could cause migration. Individually or linked together, all of the above factors can influence both individual and group decisions to migrate.

Therefore, while studying the mass migration process of the Greeks of Georgia, it would not be appropriate to rely on a single theory. In my opinion, it is desirable to use a synthesis of theories, which offers an understanding of the cause-and-effect relationships between the relevant factors in migration processes.

Neoclassical and new economics theories offer important tools for the study of the factors behind emigration and the ones influencing its first stages. By using neoclassical economic theory, we can draw out the fundamental social and economic push and pull factors that influenced the migration processes in Georgia during the 1990s and 2000s. New economics theory, for its part, offers the possibility to conduct household-level analysis, which is significant when studying the decision-making process around emigration from agriculture-dependent rural and semi-rural societies.[7] Dual/segmented labor market theory enables us to explain the demand for a low-paid workforce in receiving countries (especially in Greece and other European Union countries), as well as the economic behavior of first-generation migrants and deskilling.

World-systems theory is of no less importance. It accounts, on a structural and international level, for the dispersion of the Georgian workforce in the global labor market as a result of the country's involvement in international trade and in globalization processes.

7 Regarding international labor emigration tendencies from agriculture-dependent regions in Georgia in the 1990s and 2000s, see Zurabishvili (2007).

The factors behind Georgian Greeks' choices of emigration destinations will be explained according to this theory. Conventionally, these destinations are divided into three categories:

1. The Russian Federation, which was, at the first stage of migration, a major destination because of traditional economic and cultural links, and represented a classic example of migration flow from "periphery" to "core".
2. Greece, because of its "revitalized" economic-cultural links.
3. Western Europe and North America, which have gradually replaced the "traditional" emigration destination of Russia, because of their open borders, high wages, social guarantees, and workforce deficit. Specifically, in the case of Georgian Greeks, the fact that it was easier to get Greek, and therefore European, citizenship due to Greece's "open-door policy" in the 1990s and 2000s played an important role.

Finally, social capital theory is used to determine the characteristics of how emigration continued to develop after its first stages. Because Georgian Greeks were initially resettled in compact, often mono-ethnic villages, and due to their linguistic identity, social capital can be considered an additional incentive for emigration. These social networks offered and still offer help for potential emigrants, thus easing the emigration process.

A synthesis of the aforementioned theories should therefore be used to study the factors behind Georgian Greeks' emigration. This allows for an analysis that equally covers the micro, meso, and macro levels, and establishes individual, structural, and international driving factors for the mass emigration processes in question.

Chapter III. Greeks in Georgia: historical, spatial, and socio-cultural characteristics

3.1 Settlement history

Although the cultural relationship between Georgia and the Greek world dates back to antiquity, academic work on the history of Anatolian (Pontic and Urum) Greek settlement in Georgia has only examined these processes from the second half of the 18th century onwards. There is little information on the fate of ancient Greek settlers, or about the migrations that took place in the Byzantine era. Nor is there any high-quality academic material on their immigration into Georgian territory during the medieval or early modern periods, on the formation of their compact settlements, or on their assimilation into the local population (Kaukhchishvili 1942, Garakanidze 2000).

The first documented sources on the settlement of Anatolian Greeks in Georgia appear in 1774, linked to the development of mining activities by king Erekle II,[8] who resettled as many as 800 Anatolian Greek craftsmen and their families from the Erzurum and Kars regions to the Kingdom of Kartli-Kakheti, on the territory of modern-day Armenia (namely in Alaverdi and Akhtala, Map 2) (Mikiashvili 1989; 66). However, as a result of their geographic location and the migratory and political processes that followed their resettlement, this stream of Anatolian Greeks did not influence current Georgia's ethnic composition.

Map 2 Greek migrations to Georgia XIX-XX cc.The arrival and settlement of Pontic and Urum Greeks in Georgia took place during the 19th century and the beginning of the 20th century. This period saw the widespread resettlement, in empty or abandoned Georgian villages, of populations from the Ottoman Empire and the central territories of the Russian Empire (including Armenians, Greeks, Ossetians, Russians, Germans, Dukhobors,[9] and

8 Georgian monarch of the Bagrationi dynasty, who reigned over the eastern Georgian kingdom(s) of Kakheti (1744–1762) and Kartli and Kakheti (1762–1798).

9 Dukhobors from Russian *Духоборцы* (*Dukhobortsi*) "Spirit-Warriors" – a biblical-rational religious movement and ethnic group that appeared in Russia's Voronezh, Tambov, and Yekaterinoslav governorates. It split from the Russian Orthodox Church in the 1750–80s (Khutsishvili and Papuashvili 2008: 231–38).

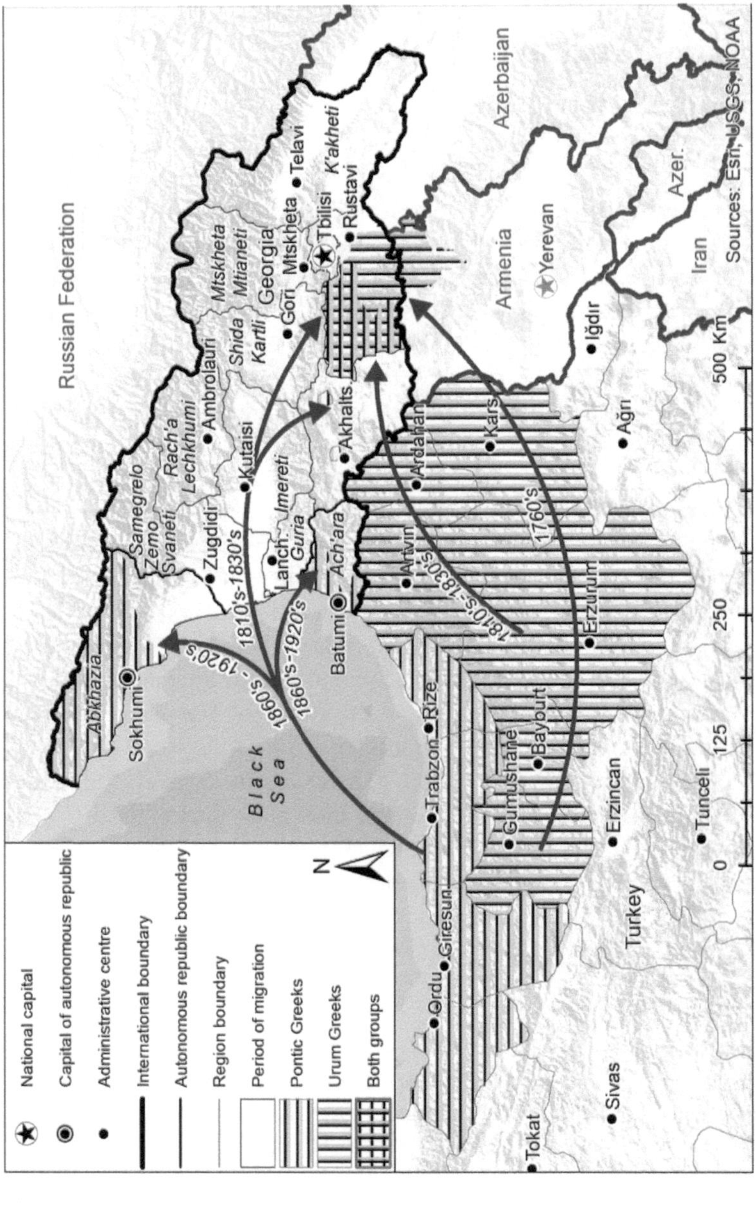

Map 2 Greek migrations to Georgia XIX-XX cc.

Molokans[10]). With this policy, the Russian Empire was trying, on the one hand, to increase the number of Christians in peripheral regions (with, for instance, Muhajirism[11] in Circassia, Abkhazia, and Adjara), and on the other hand, to develop a multiethnic society in these regions in order to hinder nationalist claims by strong local ethnic groups (Jaoshvili 1984; Gachechiladze 2008).The main causes of Greek emigration from the Ottoman Empire can be found in:

1. Military conflicts with the Russian Empire between 1820–1850;
2. The 1829 Treaty of Adrianople;
3. The 1878 Treaty of San Stephano;
4. The new economic conditions which the Russian Empire offered to Christians living in the Ottoman Empire.
5. Religious-ethnic background.

The Orthodox Greek population suffered from ethno-religious discrimination under the Islamic Ottoman Empire (Loladze 2016). The religious-ethnic identity of Greeks and Armenians became one of the Russian Empire's banners in its wars against the Ottoman Empire, and the Tsarist government often used identity to foment uprisings in the region.

Later, during the First World War (1914–18), the ascendance of Kemalism and the unsuccessful Greek military expedition to Anatolia in 1919–22 produced a new wave of refugees, a large part of whom emigrated to Greece. Others, numbering between 200,000–500,000, found refuge in the Russian Empire territories of modern-day Ukraine, Georgia, and the North Caucasus (Sideri 2006; Kitromilidis 1990; Hionidu 2012).

With the objective of religious homogenization of the old and newly conquered territories, the belligerent parties often reached agreements about population exchange. These agreements led to the aforementioned Muhajirism and the immigration of Anatolian Armenians and Greeks between the Ottoman

---

10 Molokans from Russian *Молокан* (*Molokan*), "dairy-eater" – an ecstatic-mystical as well as biblical-rational movement that split from established Dukhobor circles (Papuashvili2008: 239–45).

11 This term comes from the Arabic word مهاجر (*muhajir*), which means "displaced person" or "migrant". The term "მუჰაჯირობა" (*muhajiroba*) entered the Georgian language through Turkish, and is used to designate the forced emigration of Muslim inhabitants of the Caucasus to the Ottoman Empire precipitated by the Russian Empire in the 19th century. By displacing insubordinate Caucasian highlanders to Turkey, Russian authorities sought to impose their rule on the strategically-important region, using the vacated lands for large-scale colonization (Khorava 2004).

and the Russian Empires, as well as to the Treaty of Lausanne which required as many as 400,000 Turks to leave Greece and 1.3 million Greeks to leave the newly-established Republic of Turkey.

When discussing these processes, it is important to take into account that while the diplomatic agreements made the concerned powers responsible for immigrants' transportation and their financial and social needs, the migration was still compulsory, i.e. forced.

Maps 2, 3, 4, 5, and 6 and Annex 1 provide the locations of compact settlements. Urum Greeks mainly settled in the Kvemo Kartli region (South-Eastern Georgia), in today's Ts'alk'a, Tetrits'q'aro, Dmanisi, and Marneuli municipalities, and did not establish compact settlements in western Georgia. Pontic Greeks established compact settlements mainly in the coastal regions of Western Georgia, and specifically in the present-day autonomous republics of Adjara and Abkhazia. In addition, a relatively small number settled in the regions of Kvemo Kartli and Samtskhe-Javakheti. The historical background of these settlements and their spatial and demographic characteristics are discussed, by region, in the following sub-chapters.

3.2 Ethnic structure and ethno-demographic processes

The existing literature on demography, and general census data from the Russian Empire, the Soviet Union and independent Georgia, enable us to analyse the dynamics of Georgia's ethnic structure over two centuries. This fully covers the immigration and emigration periods of Greeks in Georgia (see Table 1, Figure 1).

Despite the fact that the first general population census in Georgia (then under the rule of the Russian Empire) took place in 1897, Soviet and modern Georgian demographers alike regard as trustworthy the available cameral census data from the 19th century and the demographic estimates made by V. Jaoshvili on their basis. They provide the best estimates of the dynamics of the ethnic structure in Georgia between 1800–1897 (Jaoshvili 1984; Jaoshvili 1996).

After 1897, the next population census only took place after the Sovietization of Georgia: in 1926, 1939,[12] 1959, 1970, 1979, and 1989. After the collapse of the

12 A general population census was also held in the Soviet Union in 1937, but the authorities at the time considered it to have been carried out with gross mistakes, and the results to be unrepresentative of the actual demographic situation. The supervisors of the census became victims of repression both in the central and peripheral regions of the Soviet Union, and a new all-Union census was held in 1939. Only in 1990 were some surviving archival materials from the 1937 census published (Tsuladze, Sulaberidze, 2015).

Soviet Union, two censuses were held, in 2002 and 2014, which covered 82 percent of the country's territory (57,000 km^2), as they could not take place in the occupied territories, the Abkhazian Autonomous Republic and the Tskhinvali Region (13,000 square kilometers in total). It is worth noting that from 1926, census data provide information about ethnic designation in both urban and rural settlements. Based on population censuses and historical processes, we can divide the ethno-demographic developments in Georgia during the last two centuries into three main phases. The first phase, from the beginning of the 19th century until Georgia's Sovietization, mainly consisted in forced migrations that led to changes in the population figures of a number of ethnic groups. The grouped settlement of non-Georgian populations in Georgia was mainly a result of Tsarist military-political and economic policies. During the second phase, from 1921 until the fall of the Soviet Union, Soviet Georgia experienced industrialization and an economic boom. Hundreds of thousands of migrants from other Soviet republics settled in Georgia. In the third phase, from Georgia's independence in 1991 to the present day, both Georgians and non-Georgians emigrated amid the political and social-economic cataclysm that followed the collapse of the Soviet Union (Komakhia 2011). Each of these phases are discussed in greater depth in the sub-chapters below.

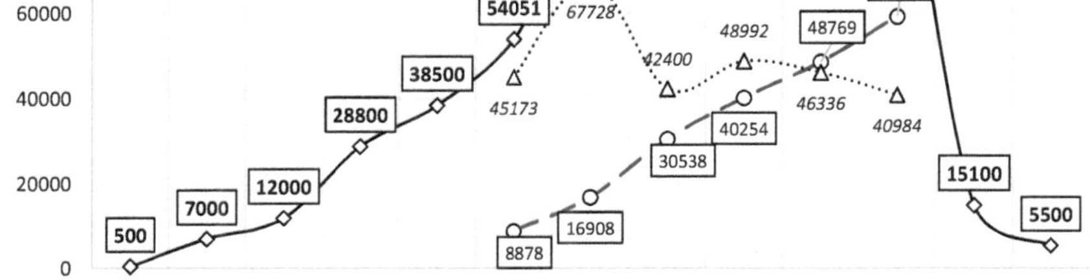

Figure 1 Dynamics of Georgia's Greek population in the years of 1800–2014

Table 1 Ethnic composition of the population in Georgia, 1800–2014

Year	Total[i]	Georg.	%	Abkh.	%	Osset.	%	Armen.	%	Russ.[ii]	%
1800[1]	784.7	622.6	*79.4*	52.0	*6.6*	29.3	*3.7*	47.0	*6.0*	-	-
1832	892.1	677.1	*75.9*	56.6	*6.3*	32.3	*3.6*	84.0	*9.4*	-	-
1865	1289.2	950.6	*73.8*	60.0	*4.6*	45.5	*3.6*	122.6	*9.5*	25.9	*2.0*
1886	1641.9	1188.1	*72.4*	38.0	*2.3*	71.0	*4.3*	172.9	*10.5*	42.5	*2.6*
1897	1919.4	1331.3	*69.4*	42.6	*2.2*	81.5	*4.2*	197.0	*10.3*	101.0	*5.3*
1926[2]	2677.2	1788.1	*66.8*	56.8	*2.1*	113.2	*4.2*	307.0	*11.5*	96.0	*3.6*
1939[3]	3540.0	2173.5	*61.4*	56.6	*1.6*	148.6	*4.2*	414.1	*11.7*	307.9	*8.7*
1959[4]	4044.0	2600.6	*64.3*	62.9	*1.5*	141.2	*3.5*	442.9	*11.0*	407.9	*10.1*
1970[5]	4686.4	3130.7	*66.8*	79.4	*1.7*	150.2	*3.2*	452.3	*9.6*	396.7	*8.5*
1979[6]	4993.2	3433.0	*68.8*	85.3	*1.7*	160.5	*3.2*	448.0	*9.0*	371.6	*7.4*
1989[7]	5400.8	3787.3	*70.1*	95.8	*1.7*	164.0	*3.0*	437.2	*8.1*	341.1	*6.3*
2002	4371.5	3661.2	*83.8*	3.5	*0.1*	38.0	*0.9*	248.9	*5.7*	67.6	*1.5*
2014[8]	3713.8	3224.6	*86.8*	>1.0	*>0.1*	14.4	*0.4*	168.1	*4.5*	26.5	*0.7*

i Note: the data from the years 2002-2014 covers the territories under Georgian jurisdiction at the time of the censuses, and therefore do not include the Autonomous Republic of Abkhazia and the Tskhinvali Region.

ii Data from the years 1800-1897 considers Ukrainians and Belarusians as Russians.

iii Data from the years 1800-1897 considers Turks and Persians as Azerbaijanis.

iv Data from the years 1939-1989 considers Yazidis as Kurds. The 2002 census counted them separately. According to the 2002 census, there were 18.3 thousand Yazidis living in territories under Georgian jurisdiction, while there where 12.2 thousand according to data from 2014.

1 For the years between 1800-1897 inclusive, I used V. Jaoshvili's calculations, which are estimates based on the first general population census of the Russian Empire in 1897 and previous cameral censuses (Jaoshvili 1984:79, Table 2.5; p 112, Table 3.8; p139 Table 4.4; p213, Table 7.1).

2 Source: All-Union Population Census (1926) *Nacional'nyj sostav naselenija po regionam respublik SSSR* (National composition of the population by regions of the republics of the USSR. Translation NL). Retrieved from http://www.demoscope.ru/weekly/ssp/sng_nac_26.php?reg=2330 (in Russian).

3 Source: All-Union Population Census (1939) *Nacional'nyj sostav naselenija po respublikam SSSR (National composition of the population by the republics of the USSR. Translation NL).* Retrieved from http://www.demoscope.ru/weekly/ssp/sng_nac_39.php?reg=4 (in Russian).

Azeri.[iii]	%	**Greeks**	%	**Jews**	%	**Germ.**	%	**Ukra.**	%	**Kurds**[iv]	%	**Other**	%
30.0	*3.8*	0.5	*0.2*	3.3	*0.4*	-	-	-	-	-	-	-	-
27.0	*3.0*	7.0	*0.8*	4.0	*0.5*	-	-	-	-	-	-	4.1	*0.5*
51.5	*4.0*	12.0	*0.9*	7.8	*0.6*	4.2	*0.3*	-	-	-	-	9.1	*0.7*
62.6	*3.8*	28.8	*1.8*	11.7	*0.7*	5.5	*0.3*	-	-	-	-	20.8	*1.3*
81.1	*5.2*	38.5	*2.0*	17.2	*0.9*	7.4	*0.4*	-	-	-	-	21.8	*1.1*
143.9	*5.4*	54.0	*2.0*	30.1	*1.1*	12.0	*0.4*	14.3	*0.5*	7.9	*0.3*	73.2	*2.8*
187.6	*5.3*	84.6	*2.4*	42.4	*1.2*	20.5	*0.5*	46.0	*1.3*	12.9	*0.3*	77.8	*2.2*
153.6	*3.8*	72.9	*1.8*	51.6	*1.3*	2.2	*>0.1*	52.2	*1.3*	16.2	*0.4*	42.0	*1.0*
217.8	*4.6*	89.2	*1.9*	55.4	*1.1*	2.3	*>0.1*	49.6	*1.1*	20.7	*0.5*	44.4	*0.9*
255.7	*5.1*	95.1	*1.9*	28.3	*0.6*	2.0	*>0.1*	45.0	*0.9*	25.7	*0.5*	45.0	*0.9*
307.5	*5.6*	100.3	*1.9*	10.3	*0.1*	1.5	*>0.1*	52.4	*0.9*	33.3	*0.6*	70.1	*1.3*
284.7	*6.5*	15.1	*0.3*	3.7	*0.1*	>1.0	*>0.1*	7.0	*0.2*	2.5	*>0.1*	38.9	*0.8*
233.0	*6.3*	5.5	*0.1*	>1.0	*>0.1*	>1.0	*>0.1*	6.0	*0.2*	>1.0	*>0.1*	34.6	*0.9*

4 Source: All-Union Population Census (1959) *Nacional'nyj sostav naselenija po respublikam SSSR (National composition of the population by the republics of the USSR. Translation NL).* Retrieved from http://www.demoscope.ru/weekly/ssp/sng_nac_59.php?reg=8 (in Russian).

5 Source: All-Union Population Census (1970) *Nacional'nyj sostav naselenija po respublikam SSSR (National composition of the population by the republics of the USSR. Translation NL).* Retrieved from http://www.demoscope.ru/weekly/ssp/sng_nac_70.php?reg=6 (in Russian).

6 Source: All-Union Population Census (1979) *Nacional'nyj sostav naselenija po respublikam SSSR (National composition of the population by the republics of the USSR. Translation NL).* URL: http://www.demoscope.ru/weekly/ssp/sng_nac_79.php?reg=6 (in Russian).

7 Source: All-Union Population Census (1989 *Nacional'nyj sostav naselenija po respublikam SSSR (National composition of the population by the republics of the USSR. Translation NL).* Retrieved from http://www.demoscope.ru/weekly/ssp/sng_nac_89.php?reg=6 (in Russian).

8 Source: Geostat (2016) *Statistical Yearbook of Georgia.* Tbilisi: National Statistics Office of Georgia: 22

3.2.1 Russian Empire

The 19th-century cameral and general population censuses did not include any question about ethnic belonging. However, the 14 questions included ones about place of birth, religious confession, and mother tongue. When combined with other sources (for instance documentation about migrations, military operations, and groups of colonists), demographers can make reasonable estimates of the ethnic structure of Georgia.

As Table 1 shows, the 19th century saw important changes in the ethnic structure of Georgia's population. The percentage of ethnic Georgians dropped from 79.4% in 1800 to 69.5% in 1897. In parallel, the number of immigrant Armenians, Russians, Germans and Greeks increased.

In 1800, before the mass migrations, the number of Greeks in Georgia was only about 500 (0.06% of the total population). Over the following century, their numbers increased 77-fold to 38.5 thousand in 1897 (2% of the total population). This rapid increase was obviously the result of migration linked to the Russo-Turkish conflict and the Russian Empire's ethnic policy in Georgia, as discussed in sub-chapter 3.1.

3.2.2 The Soviet Union

As the 1926 and 1939 censuses show, large migration processes continued after Georgia became part of the Soviet Union, and until the Second World War. As a result, in 1939 ethnic Georgians made up 61.4% of the population, for the first time less than two thirds of the Georgian Soviet Republic's total population. As for ethnic minorities, Russians and Ukrainians experienced the most notable increases, again a consequence of mass immigration processes in the 1920s and 1930s (see Table 1).

Large-scale industrialization in Georgia began in the 1920s and attracted many migrants from other Soviet republics. This immigration flow also stemmed from an increased demand for highly qualified professionals, due to the rapid development of industry and agriculture. Additionally, as the collectivization and "de-kulakization" campaigns (1930–1933) took the lives of several millions in southern Russia, Ukraine, and Kazakhstan, the republics of the southern Caucasus sheltered hundreds of thousands of people fleeing famine (mostly Russians and Ukrainians). Their numbers increased by 320% between 1926 (110,500 individuals) and 1939 (354,000 individuals) (Komakhia 2011; Gachechiladze 1997).

The Greek population of Georgia also grew rapidly between 1897–1939. According to census data, during these 42 years the number of Greeks increased by 119.8%, from 38,500 to 84,636. This increase was partly due to a high natural growth rate (Komakhia 2011;10). Immigration from the Ottoman Empire also played an important role until the end of the 1920s, when the border was closed following the end of the conflict in Anatolia and the Soviet occupation of Georgia.

The dynamics of Greek population numbers in Georgia vary between urban and rural settlement types. As Figure 1 shows, in 1926, 83.5% (45,000) of the Greeks in Georgia were living in villages and only 16.5% (8,800) in urban areas. Despite small changes, this significant difference remains evident in the 1939 census data: 80% (67,728) of the Greek population in Georgia were living in rural areas and 20% (16,908) in urban ones. This imbalance was due to the organized character of the immigration, as mentioned in the previous sub-chapter, within the context of the successive regimes' spatial ethnic policies. The reduction, which persisted in subsequent years, can be attributed to the most widespread form of internal migration: from villages to cities. The intense industrialization process taking place in Georgia from the 1930s resulted in a deficit of workers in cities, encouraging rural-urban migration.

Unfortunately, after 1939, due to the Second World War and its aftermath, no population censuses were carried out until 1959. Consequently, there is no accurate data about the demographic situation of that period. Even though the war did not extend to Georgia, its absolute population size decreased due to the total mobilization policy and the great losses endured in combat.[13] From 1946 on, the population size steadily increased. Pre-war population numbers are thought to have been reached again only in 1955 (Jaoshvili 1996; Jaoshvili 1944). [14] From the end of the 1930s until the 1950s, important changes took place in the ethnic structure of the Georgian population. These came about primarily due to the mass repression and deportations carried out in the border regions of the Soviet Union. These were mostly of an ethnic nature. Because these tragic events resulted in a significant decline in the population size of a number of ethnic groups in Georgia, and especially of Germans and Greeks, I will discuss them in detail in Chapter IV.

---

13 The commonly accepted estimate in Georgian historiography is that up to 700,000 people from Georgia were mobilized in the Army during the Second World War, which was more than 18% of the total Georgian population at that time. It is believed that the number of casualties in battle reached up to 300,000 individuals (8% of the total population) (Babalashvili 1977 p 318; Jaoshvili 1996 p 152). That said, there are no accessible primary sources to verify these numbers.

14 Because no census was held in 1940–1958, this is an estimate from the statistics services of the time. According to V. Jaoshvili, the population size could not have reached pre-war numbers that fast (Jaoshvili 1997; pp 150–161).

The data from the 1959 census provide interesting material for the study of the Georgian population's ethnic structure. According to this census, it was the first time in 150 years that ethnic Georgians had increased their share in the population (by 2.9%). This increase should be attributed to both natural and mechanical movements – the post-war increase in birth rates, and the decrease of immigration flows since the 1930s (Jaoshvili 1996; Natmeladze 2002).

A new phase of demographic development in Georgia started in the 1960s and lasted until 1989. This period is characterized by a gradual reduction of natural population growth and a negative migration balance. For instance, in the 1950s the yearly population increase was 1.63%. The same indicator was 1.35% in the 1960s, and 0.75% in the 1970s – less than half as much as in the 1950s.

These changes resulted both from the reduction in birth rates and an increase in death rates. This can be considered a "demographic echo" to the Second World War, as the generation born during the Second World War became of marriageable age, and their relative number in the total demographic structure was significantly lower than normal (Jaoshvili 1996:157–158).

Despite a reduced birth rate, the proportion of ethnic Georgians increased steadily by 2–3% in Soviet Georgia between 1960–1989. This was due to the change in the mechanical movement vector of ethnic minorities.

The principal vectors, nature, and scale of migration gradually changed from the second half of the last century. Until then, resettlement of the Slavic (mainly Russian) population in the peripheries of the USSR slowly took place. In the 1960s, the opposite occurred. In the 1970s, the outflow of the Russian-speaking population from non-Slavic republics reached a mass scale. As a result, in the 1980s, a negative migration balance of the Russophonic population became characteristic for all the non-Slavic republics of the USSR (Janiashvili 2008).

Apart from internal migration, Jewish emigration to Israel was permitted in 1971, and led to an almost five-fold drop in the Georgian Jewish population between 1970 (55,400) and 1989 (10,300).

During the second half of the 20th century, the size of the Azerbaijani population in Georgia increased faster than all its other ethnic groups, doubling over 30 years from 153,600 Azerbaijanis in 1959 (3.8% of the total population) to 255,700 in 1979 (5.1%) and 307,600 in 1989 (5.7%).

As a result of the repression and deportations carried out in the 1940s and 1950s, the Greek population declined from 84,400 to 72,900 (almost 14%). There were no other significant compulsory or voluntary mechanical movements. As a result, numbers did not decrease until the collapse of the Soviet Union. Instead, as a consequence of natural growth, they increased at a slow and steady rate, reaching 89,000 (1.9%) in 1970, 95,000 (1.9%) in 1979, and a historic high of 100,324 (1.9%) in 1989.

3.2.2.1 Trends in the rural-urban distribution

The censuses carried out under the Soviet Union are the only ones that provide sufficient information about the distribution of Greeks between urban and rural settlements. The data is presented in Figures 2 and 3. It enables us to compare the type of Greek settlements to those of the Soviet Georgian territory as a whole.

As noted above, after the Stalinist repression and deportations, which mainly affected Greeks living in compact rural settlements in western Georgia, the number of Greeks living in villages in 1959 was 42,400 – a decrease of 21.9% since the 1939 census (see Figure 2). The parallel increase in the urban population reduced the ratio further. In 1959, the percentage of Greeks living in rural areas of Georgia was 58.1% (42,400), while 41.9% (30,538) were living in urban areas.

The increase in urban population continued until the collapse of the Soviet Union. Greeks in urban settlements carried the balance for the first time in the 1979 census, where the ratio was 51.3/48.7. The 1989 census shows an even larger difference: Greeks living in cities constituted 59.1% (59,340) of the total Greek population in Georgia and those living in villages, 40.9% (40,984) (see Figures 2–3).

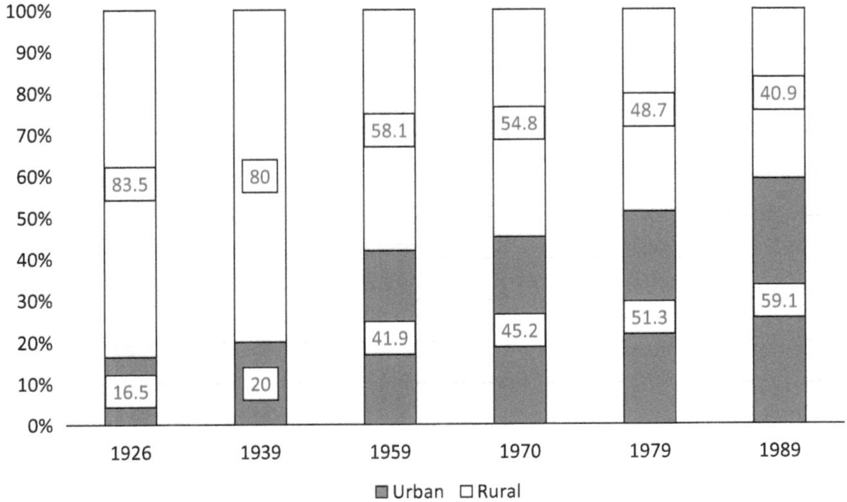

Figure 2 Distribution of Greeks of Georgia in relation to settlement type, according to general population census from 1926–1989

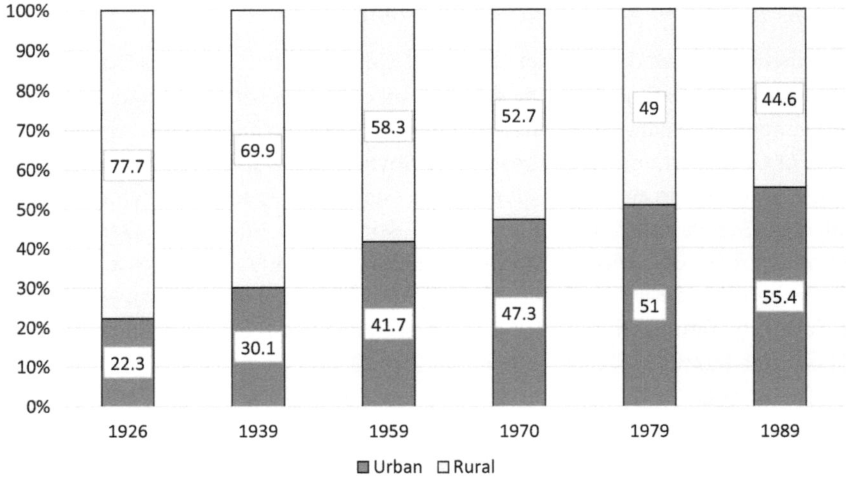

Figure 3 Distribution of total population in Georgia in relation to settlement type, according to general population census from 1926–1989

All in all, the development of the Greek population does not reveal anything special about this ethnic group. As seen above, the trend in the rural-urban distribution is almost identical to that of Soviet Georgia as a whole. This suggests that Greeks were a well-integrated group in Soviet Georgia, and did not deviate from the general demographic characteristics of the country.

3.2.3 Independent Georgia

In the 1990s, the new reality that followed the collapse of the Soviet Union, which in the first place meant the fall of the Iron Curtain and more freedom of movement, had direct repercussions on migration processes in Georgia. As soon as it declared independence, the country experienced an economic collapse, record inflation and unemployment, and the collapse of state institutions. The nationalist surge that followed 70 years of Soviet occupation dragged the country into ethnic conflicts, which were important triggers for the activation of emigration processes. The emigrants mainly left for their ancestral homelands, to various cities in the Russian Federation, and to western European countries (Loladze 2016; CRRC 2007).

5.4 million people lived in Georgia in 1989. According to official data, by 2007 the population had declined by 19% to 4.3 million people. In 2014, with 3.7 million people, it had shrunk by 31% compared to 1989, and by 15% compared to 2002.

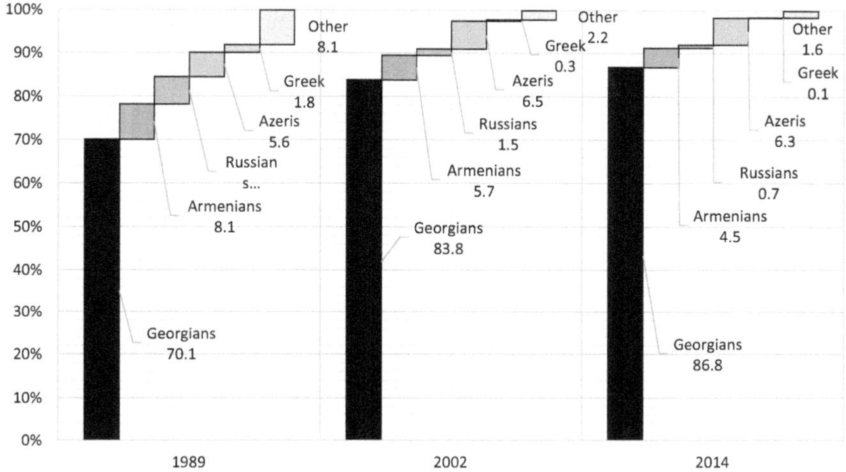

Figure 4 Distribution (%) of population by ethnicity in Georgia, 1989–2014

Among the migration processes that began in 1990, the most intense was the emigration of Slavs (ethnic Russians, Ukrainians, and Belarusians). Many of them were members of the Soviet military, or civil servants, who with their families emigrated *en mass* from Georgia following the dislocation of military units. Among these emigrants, more than 60% were Slavs. Most of them (88.5%) were settled in cities, and almost all emigrated to cities in the Russian Federation.

According to the results of the 2002 Georgian population census, the size of the Russian population had declined significantly since the last Soviet census of 1989, from 341,000 in 1989 (6.3% of the total population) to only 67,671 (1.5%) in 2002. This trend continued, and in the 2014 census data the number of Russians in Georgia was down to 26,500 (0.7% of the total population).

Ossetians were one of the largest ethnic minority groups in Georgia. At the end of the 1980s, there were 164,000 Ossetians (3% of the total population) on Georgian territory, of which up to 65,000 lived in the South Ossetian Autonomous Oblast. During the Georgian-Ossetian conflict,[15] some Ossetians moved and settled in the Russian Federation. As per the 2002 census, there were 38,000 Ossetians in Georgia, outside the occupied territories, and in 2014 there were

15 This refers to the 1991–92, armed conflicts, in which Georgian authorities lost access to Tskhinvali and its neighboring villages, which were mostly inhabited by ethnic Ossetians.

14,400. Similarly to Ossetians, the significant reduction of the Abkhaz population of Georgia in the censuses is directly linked to military operations and separatist conflicts. According to the 1989 census, 93.300 out of the 95,900 Abkhazians living in Georgia were settled in Abkhazia. Since the Georgian government had lost control over this region as a result of the military conflict that took place in 1993–94, the 2002 and 2014 censuses did not include Abkhazia. The data from the census carried out by Abkhazia's *de facto* government is discussed in sub-chapter 3.5.

The harsh socio-economic and political processes of the 1990s had a direct effect on Greek migration behavior, and in the following decades resulted in a dramatic 95% decline in the Greek population in Georgia. Numbers fell from 100,324 Greeks in 1989 (1.9% of the total population) to 15,100 (0.3%) in 2002 and 5,500 (0.1%) in 2014.

Finally, as seen in Figure 4 and Table 1, the population declines affected every group (as a result of the socio-economic conditions which are discussed in Chapter V). However, the impact is more severe with non-Georgian groups. This is subject to further analysis below, as it had a crucial repercussion: the transformation of Georgia from a multiethnic state towards being a monoethnic one.

3.3 Kvemo Kartli

The organized and compact settlement of Pontic and Urum Greeks in Georgia began in the Kvemo Kartli region in the end of the 1820s.[16] Earlier, in 1806–1813, small groups of Greek refugees from the Ottoman provinces (*Eyalet*) of Kars and Erzurum had already settled individually in the villages of Ts'ints'q'aro and Iraga (both in the Tetrits'q'aro Municipality). Some also settled independently in Tbilisi, Batumi, Poti, and other cities, from the first half of the 19th century until the 1920s (Pashaeva 1992; Melikishvili 2011; Loladze 2016).

Subsequently, organized resettlement took place in three distinct stages. The first, in September 1829, involved 60 Greek families from Gümüşhane and Maden in present-day Turkey, later joined by 85 families from the villages of Kars and Bayburt. They found a temporary refuge in the village of Ts'ints'q'aro among Greeks who had settled there earlier. In 1830, however, due to the lack of sufficient agricultural land in Ts'ints'q'aro, the Greeks from Gümüşhane, Maden, and Bayburt sought to improve their living conditions by moving to the village

16 L. Pashaeva and other authors distinguish four separate stages of migration, but the fourth stage that they refer to only concerned the Georgian population of Ardagani, which settled in the village Rekha of the Ts'alk'a Municipality.

of Beshtasheni in the Ts'alk'a Municipality. This became the first permanent settlement of Greeks in Georgia (Pashaeva 1992).

In the second stage, a group of Anatolian Greeks came in June 1830 to Kvemo Kartli from the village Bashkev in the present-day Erzurum province. The group settled near Ts'alk'a, in the abandoned village of Zemo Akhlasheni, and would later change its name to Bashk'ev-Kendig.

The third stage of migration similarly originated from a number of Sanjaks (districts) in the Erzurum Eyalet, continuing up to the autumn of 1830. The migrants again settled in a number of abandoned villages near Ts'alk'a. As a result, by the end of 1830 there were already 18 Greek villages in the Ts'alk'a area (Kauchkhishvili 1942).

In 1829–30, apart from Tetrits'q'aro and Ts'alk'a, Greek settlements appeared in the present-day Dmanisi Municipality. The Urum who moved to this territory mainly came from the northeastern Vilayet (administrative divisions) of the Ottoman Empire. They eventually took up residence in three villages: Ganakhleba, Velisp'iri, and Sark'ineti.

As a result of the abovementioned processes, compact Greek settlements mainly appeared in Kvemo Kartli, in present-day Tetrits'q'aro and Ts'alk'a municipalities, and to a smaller extent, in villages in the Dmanisi, Bolnisi, and Marneuli municipalities. Of the 41 Greek villages in Kvemo Kartli, 30 were mostly inhabited by Urum Greeks and 11 by Pontic Greeks (see Map 3, Appendix 1.)

Compared to Adjara, Abkhazia, and Georgia as a whole, the materials from the general and cameral censuses from the 19th century provide less detailed information about ethnic processes in the Kvemo Kartli and Samtskhe-Javakheti regions. According to the National Statistics Office of Georgia, the lack of detailed information about the ethnic composition of these regions at the time is due to changes in their administrative borders. This it necessary to carry out additional research in order to calculate precise numbers.

From 1926, when the ethnicity indicator was introduced into the census questionnaire, it is possible to analyze the ethnic processes of these regions at the municipal level. This enables the reconstruction of these processes using modern borders (see Table 2–3).[17]

As mentioned above, at the end of the 1920s in Georgia, and therefore in Kvemo Kartli, large foreign immigration processes ended together with the consolidation of Soviet rule. As a result, the ethnic structure of the population mainly

17 These calculations were only necessary for the 1926 and 1939 data. Data from the 1959–2014 censuses were provided to by the National Statistics Office of Georgia.

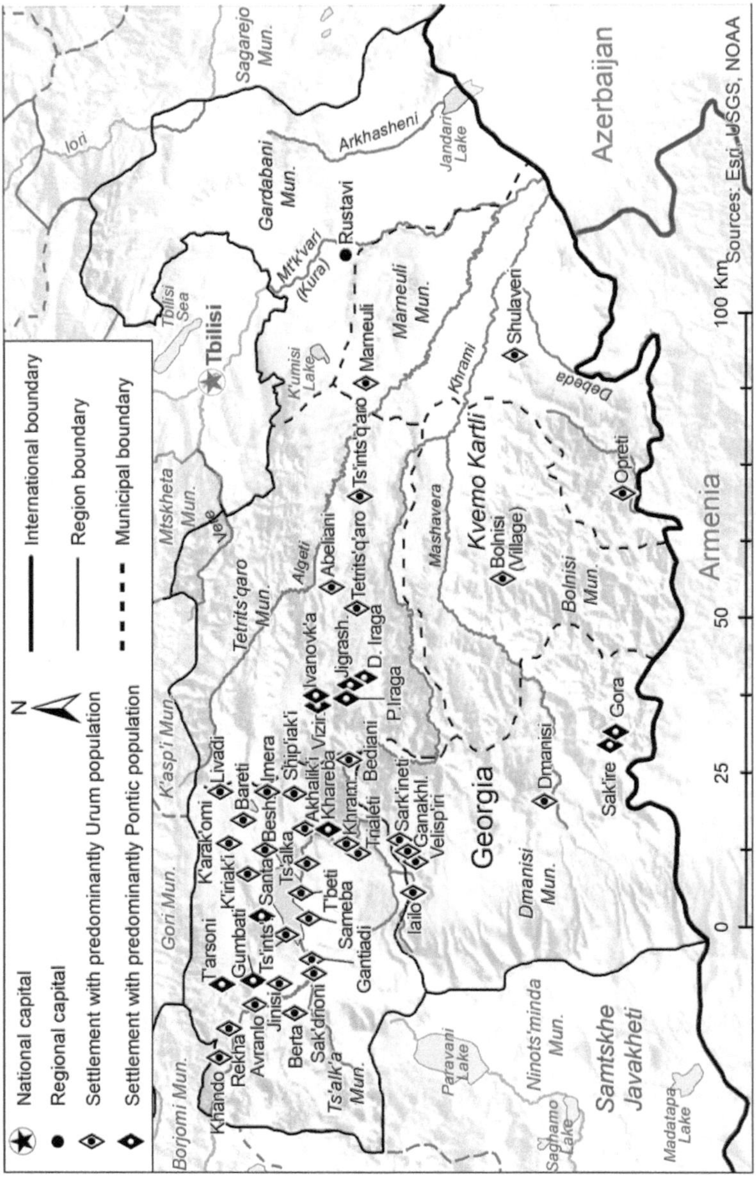

Map 3 (Former) Greek compact settlements in Kvemo Kartli

Table 2 Ethnic composition of the population in Kvemo Kartli in 1926–2014 according to the data used in Table 1

	Total	**Georgian**	%	**Ossetian**	%	**Russian**	%	**Azeri**	%	**Arme-nian**	%	**Greek**	%	**Other**	%
1926	181 222	26 260	*14.5*	2 208	*1.2*	7 713	*4.2*	65 851	*36.3*	37 858	*20.8*	30 065	*16.5*	11 267	*6.2*
1939	219 603	28 245	*12.9*	3 262	*1.5*	14 691	*6.7*	81 364	*37.0*	44 155	*20.1*	34 898	*15.9*	12 988	*5.9*
1959	298 493	64 729	*21.6*	4 856	*1.6*	30 013	*10*	99 655	*33.3*	48 004	*16.0*	43 943	*14.7*	7 256	*3.6*
1970	498 573	154 920	*31.0*	7 303	*1.4*	41 606	*8.3*	177 948	*35.6*	53 173	*10.6*	52 951	*10.6*	3 963	*1.6*
1979	528 468	167 760	*31.7*	7 976	*1.5*	37 768	*7.1*	203 009	*38.4*	49 530	*9.3*	52 186	*9.8*	5 692	*1.8*
1989	608 491	214 347	*35.2*	8 587	*1.4*	34 934	*5.7*	243 975	*40.1*	43 993	*7.2*	50 252	*8.2*	7 433	*2.0*
2002	497 530	222 450	*44.7*	2 184	*0.4*	6 464	*1.3*	224 606	*45.1*	31 777	*6.3*	7 415	*1.4*	1 924	*0.4*
2014	423 986	217 305	*51.2*	810	*0.1*	2 631	*0.6*	177 032	*41.7*	21 500	*5.0*	2 113	*0.5*	2 063	*0.5*

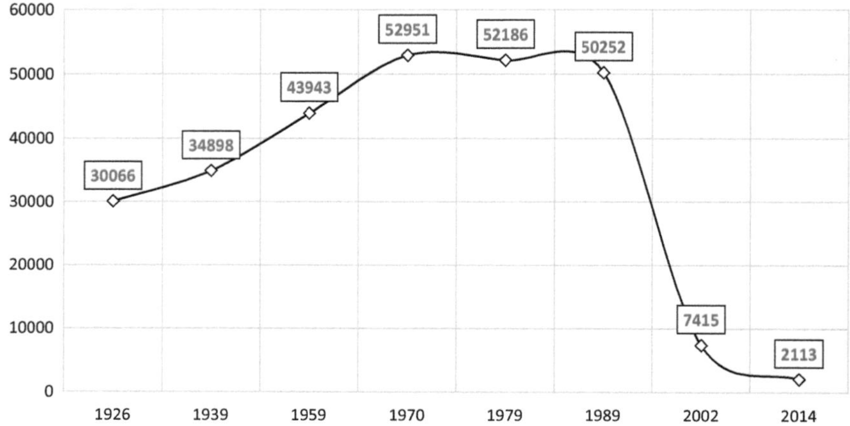

Figure 5 Dynamics of the Greek population in Kvemo Kartli in 1926–2014 according to the data from Table 2

changed due to natural movement, a relatively smaller scale of immigration compared to the previous century, and to the later forced migrations of the 1940s.

The ethnic policies of the Russian Empire in the 19th century transformed Kvemo Kartli and Samtskhe-Javakheti into multiethnic regions. This remains the case to this day. As Table 2 shows, in 1926 Azerbaijanis were the major ethnicity (36.3% of the total population of the region). They remained so until 2014. The second largest ethnic group in 1926 was Armenians (20.8%), followed by Greeks (16.5%) and Georgians (14.55%). This order remained in

the 1939 census, which showed an important increase only in the number of Russians. Their population had almost doubled due to immigration, from 7,713 to 14,691. This order was the same according to the 1939 census, which showed an important increase only in the number of Russians. Their population had almost doubled due to immigration, increasing from 7 713 to 14 691.

Important changes in the region's ethnic structure appear in the 1959 census, when as a result of mechanical movement, the number of Georgians in the region increased by 129% compared to 1939, from 28,000 to almost 65,000. In addition, the number of Russians rose from 14,000 to 30,000, an increase of 104%. The share of "Others" declined by 44%. The number of Greeks and those of other ethnic groups represented in Table 2 changed "only" as a result of natural movement. Therefore, these ethnic groups did not see any important increase or decrease in terms of population.

The changes that took place in the region between 1939–1959 were due to two main factors:

1. The significant increase of the Georgian and Russian populations was a result of the construction of a large Soviet industrial city in Rustavi in 1948, with a large metallurgic industrial complex and other large and medium factories. This remained an important industrial center of the Transcaucasia region until the 1990s.
2. In 1937–1952, Soviet authorities carried out repression and deportations on ethnic grounds in Kvemo Kartli. In this region, the main targets of the Stalinist regime were compactly settled Germans. The Greeks of Kvemo Kartli, however, unlike those in Abkhazia and Adjara, did not suffer from repression (see Chapter IV).

According to the 1970 census, the significant increase in the number of Georgians and Azerbaijanis continued in the region. This was again linked to the development of Rustavi and to high birth rates among the Azerbaijani population. Compared to data from the 1959 census, the ethnic Georgian population increased by 139%, from 64,000 to almost 155,000, representing 31% of the region's total population. The number of Azerbaijanis increased by 78%, reaching 177,000 and making up 35.6% of the region's total population.

The remarkable increase in the Georgian and Azerbaijani populations resulted in a decrease in the percentage of the other ethnic groups, even though their numbers were still increasing due to significant migration processes and natural growth.

In the 1970s, like in the rest of Georgia, natural population growth in Kvemo Kartli diminished and mortality rose. For the first time there was also a negative migration balance. The emigration of the Russian-speaking[18] population was particularly noticeable. Together with the aforementioned factors, it led to stagnation in the region's population dynamics. The 1979 census was the first to show a decrease in the number of Russians (-9.2%), Armenians (-6.8%) and Greeks (-1.4%) in the region. The rise in emigration, and the decline of immigration linked to Rustavi, also led to a slow-down in the growth of the Georgian and Azerbaijani populations in the region. For instance, by 1979 the percentage of Georgians in the total Kvemo Kartli population had barely changed. Ethnic Georgians made up 31.7% of the population, having grown by only 8.2%, from 154,000 to 167,000. Compared to Georgians, the Azerbaijani population had a higher growth rate of

18 By Russian-speaking population, we can encompass not only Slavs, but also all the other ethnic minorities of the region. Because the Russification policy started in the 1930s, lessons were provided in Russian languange in the schools of compact settlements of ethnic minorities (Höfler 2020;33–36).

14%. Still, this was radically lower than in previous years. Overall, in 1979 there were 203,000 Azerbaijanis in the region, making up 38.4% of its total population.

This stagnation until the collapse of the Soviet Union, as seen in the 1989 census data. Again, Georgian and Azerbaijani populations were the exceptions in this regard. The birth rate among Azerbaijanis in Kvemo Kartli was typically high, with their population growing by 20% between 1979–1989, from 203,000 to 243,000.

Environmental migration during this period had a direct influence on the growth of the Georgian population of the region. In 1987, heavy snowfalls in the mountain region of Svaneti required a mass evacuation and resettlement of the population. Following a governmental order, 3,000 families were moved from Svaneti to the Tetrits'q'aro district during 1988–1989, and six new villages were built for them. Then, in 1989, landslides in the Adjara region led to 85 additional families being resettled in Tetrits'q'aro. These events, coupled with higher natural population growth, raised the number of Georgians in Kvemo Kartli to 214,000 in 1989, 35.2% of the total population. The same census records the continuing decline of the Greek (-3.8%), Armenian (-12%) and Russian (-8%) populations.

The processes that unfolded after the collapse of the Soviet Union led to an overall decline in the region's population. Emigration took on a mass scale and concerned the entire population. According to the 2014 census, the total population of the region decreased by 43% compared to 1989, from 608,000 to almost 424,000. As a result, the region's ethnic structure was also transformed. The percentage of Georgians increased to 51.2% (217,000 individuals), while the number of Azerbaijanis decreased by 27% and reached 177,000, making the ethnic Georgian population the largest ethnic group in the region for the first time. The Armenian population was nearly halved, now making up 5% of the region's population (21,500 individuals).

The challenges that arose when Georgia gained its independence, discussed in Section 3.1, also radically reduced the ethnic diversity of Kvemo Kartli's population, which had developed over the course of two centuries. The region was practically emptied of Ossetian, Russian, and Greek populations. At present, these groups together make up only 1.1% of the region's total population (see Table 2).

3.4 Samtskhe-Javakheti

The territory of the present-day Samtskhe-Javakheti region holds the smallest number of the Greek settlements considered in this study. Like Kvemo Kartli, Greek refugees moved there from the beginning of the 19th century until the 1860s.

The first wave came from the northeastern territories of the Ottoman Empire in the 1820–30s, and settled in Akhalkalaki, among the existing ethnic Armenian population. Over time, the immigrants assimilated into the local Armenian population. Later, in the 1860s, Pontic Greeks founded two villages: Mikelts'minda, in the present-day Akhaltsikhe Municipality, and Tsikhisjvari, in the Borjomi Municipality. Pontic Greeks also lived alongside Georgians and Armenians near Tsikhisjvari, in the village of Andezit'i. Tsikhisjvari is completely isolated from the rest of the Greek settlements in Georgia, and has remained a compact, monoethnic Greek settlement to this day (see Map 4) (Topchishvili 2015, Pashaeva 1983, Loladze 2016).

Regarding the Samtskhe-Javakheti region, Table 3 shows that from 1926 to the end of 2014, Armenians were the largest ethnic group in the region. Greeks, in contrast to the rest of the regions where they were compactly settled, were the smallest ethnic group in Samtskhe-Javakheti. Their demographic behavior did not play a noteworthy role in the region's ethnic structure (see Picture 4).

From the 1920s, like in Kvemo Kartli, immigration from the Ottoman Empire to Samtskhe-Javakheti had ceased. Hence, the Greek population only saw natural growth. According to census data, it peaked in 1959 with 1,825 individuals. This corresponded to 0.6% of the total population of the region. Data from the censuses between 1970–1989 show the Greek population in Samtskhe-Javakheti declining at a slow rate, due to mixed marriages and rural-to-urban emigration in the region.

As with Kvemo Kartli, the Soviet repression carried out in the years 1930–50 did not affect the Greeks living in Samtskhe-Javakheti, as it targeted the region's Muslim population[19] (affecting its vast majority), to whom the census data refers as Azerbaijanis (see Table 3).

From 1959 to 1989, as in Kvemo Kartli and the rest of Georgia, emigration from Samtskhe-Javakheti increased and the birth rate dropped. This resulted in the stagnation and then a slight decline of the population size. In the period following Georgia's independence, as in other regions, emigration from Samtskhe-Javakheti increased, decreasing the population of all ethnic groups in the region and thus its total population. According to 2014 data, there were 420 Greeks in the region, most of them living in the only remaining compact Greek settlement, Tsikhisjvari (see Table 3, Map 4).

19 Until the Second World War, Meskheti's Muslim population had various collective appellations: "Tatars", "Azerbaijanis", then "Meskhetian Turks" or "Turk-Meskhetians". They were called "Akhaltsikhian Turks" (Turkish: Ahıska Türkleri) in Turkey. According to professor Revaz Tabukashvili, all these appellations are incorrect, as in reality this was a conglomerate of ethnic groups (Islamized Georgians, various Turkish tribes, Kurds, and Hemshin – Islamized Armenians) who only had Sunni Islam in common. Tabukashvili supports this view with data showing that among those deported from Samtskhe-Javakheti, 70% considered themselves to be Turks, 16% Azerbaijanis, 5% Kurds, and the rest – Tatars or Yezidis (Gachechiladze 2008: 446–7).

Table 3 Ethnic composition of the population in Samtskhe-Javakheti in 1926–2014 according to the data used in Table 1

	Total	Georgian	%	Ossetian	%	Russian	%	Azeri[1]	%	Armenian	%	Greek	%	Other	%
1926	196 854	45 243	*22.9*	3729	*1.8*	5 015	*2.5*	56 117	*28.5*	75 402	*38.3*	1 292	*0.6*	10 056	*5.1*
1939	266 115	44 941	*16.8*	5116	*1.9*	13 503	*5.0*	88 164	*33.1*	105 440	*39.6*	1 695	*0.6*	2 275	*0.8*
1959	213 329	70 381	*32.9*	3894	*1.8*	14 766	*6.9*	189	…	118 431	*55.5*	1 825	*0.8*	5 668	*2.6*
1970	231 593	79 225	*34.2*	4104	*1.7*	12 586	*5.4*	307	*0.1*	129 898	*56.0*	1 815	*0.7*	5 473	*2.3*
1979	226 051	78 148	*34.5*	3999	*1.7*	10 639	*4.7*	472	*0.2*	128 574	*56.8*	1 771	*0.7*	4 219	*1.8*
1989	235 512	87 408	*37.1*	3572	*1.5*	10 431	*4.4*	454	*0.1*	128 204	*54.4*	1 693	*0.7*	5 443	*2.3*
2002	207 598	89 995	*43.3*	822	*0.3*	2 230	*1.0*	59	…	113 347	*54.5*	740	*0.3*	1 145	*0.5*
2014	160 504	77 498	*48.2*	393	*0.2*	712	*0.4*	89	…	81 089	*50.5*	420	*0.2*	723	*0.4*

1 according to the 1939 all union census, all turkophonic muslim population in Georgia were recorded as Azerbaijanis. In case of Javakheti region, their absolut majority where "meskhetian turks (Gachechiladze 2008; 304)

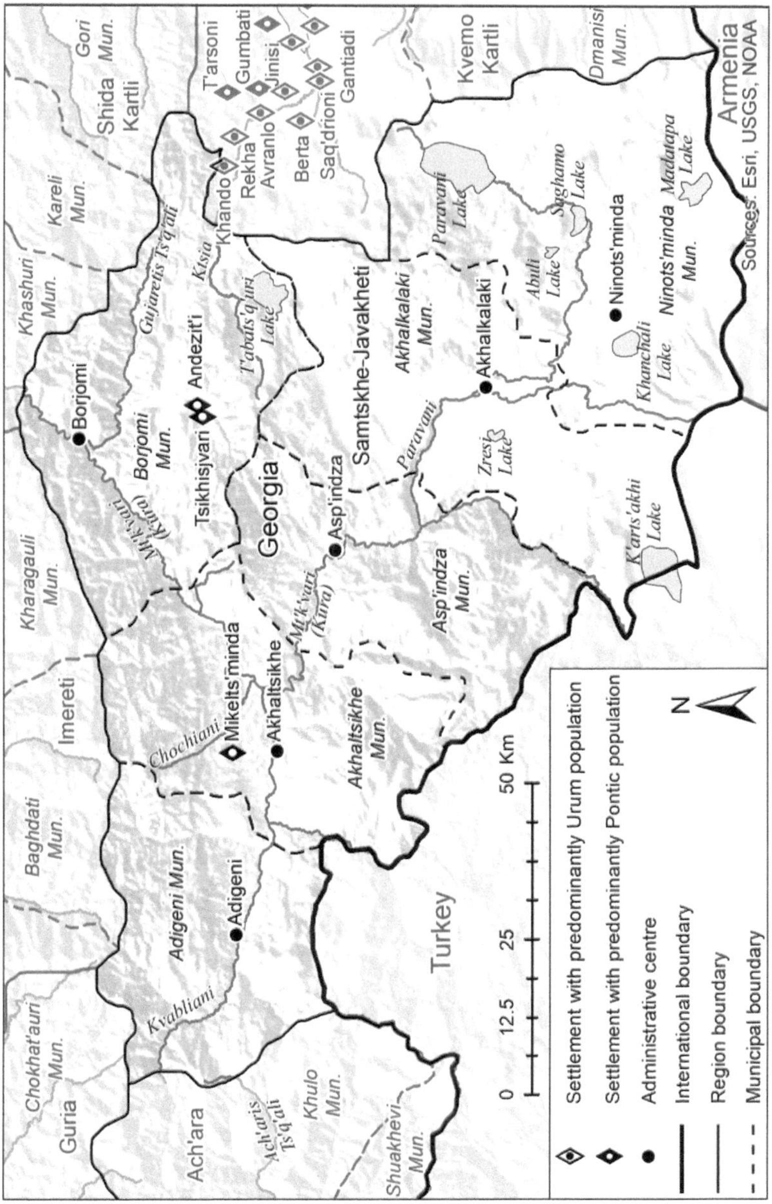

Map 4 (Former) Greek compact settlements in Samtskhe-Javakheti

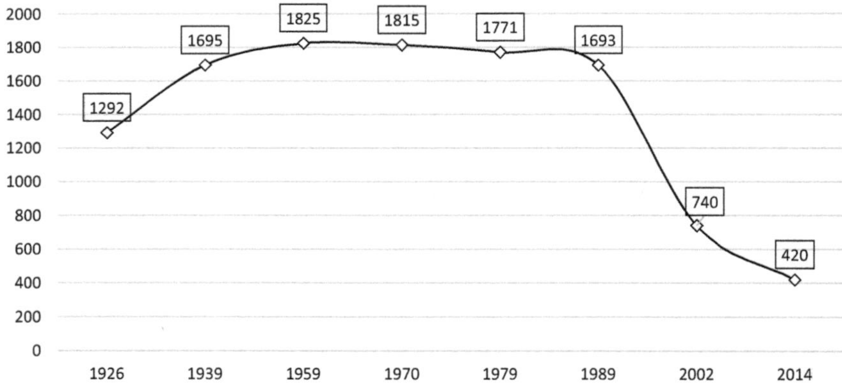

Figure 6 Dynamics of Greek population in Samtskhe-Javakheti, 1926–2014

3.5 Abkhazia

The resettlement of Greeks in northwestern Georgia, on the Abkhazian Black Sea coast, is directly linked to the Abkhaz revolt of 1866 and its consequences, Muhajirism and the Russo-Turkish war of 1877–78 (Jaoshvili 1996).

In 1866, Russia attempted to annex the Abkhazian principality and to establish Imperial rule and administration in the region. The resultant discontent among the local population led to a revolt breaking out later the same year, which was soon quelled, As a consequence, the Russian Empire decided to implement a radical ethnic policy in the region to establish its rule, and to replace the Muslim population which it considered untrustworthy with colonists who would be more loyal.

The mass deportations of the local population from Abkhazia were thus perpetrated on ethnic and religious grounds, with 15,000 Muslim Abkhazian *muhajirun* forced to move and resettle on Ottoman territory. Further deportations occurred after the Russo-Turkish war of 1877–78, when the Russian Empire restored the rule it had temporarily lost in Abkhazia. According to V. Jaoshvili's data, at least another 15,000 Abkhazians were forced to leave, leading to an overall population decrease of 12.9% in the region, with the ethnic Abkhazian population nearly halved (Jaoshvili, 1968. Jaoshvili 1984:90–91).

From the 1860s, during and after Muhajirism, a mass immigration process began into the Abkhazian territory. A large part of the immigrants were resettled Armenians and Greeks coming from Turkey, through the Black Sea, who mainly settled in the present-day Sokhumi, Gulripshi, and Gagra municipalities (see Map 5). Russians and Estonians also moved to Abkhazia during the same

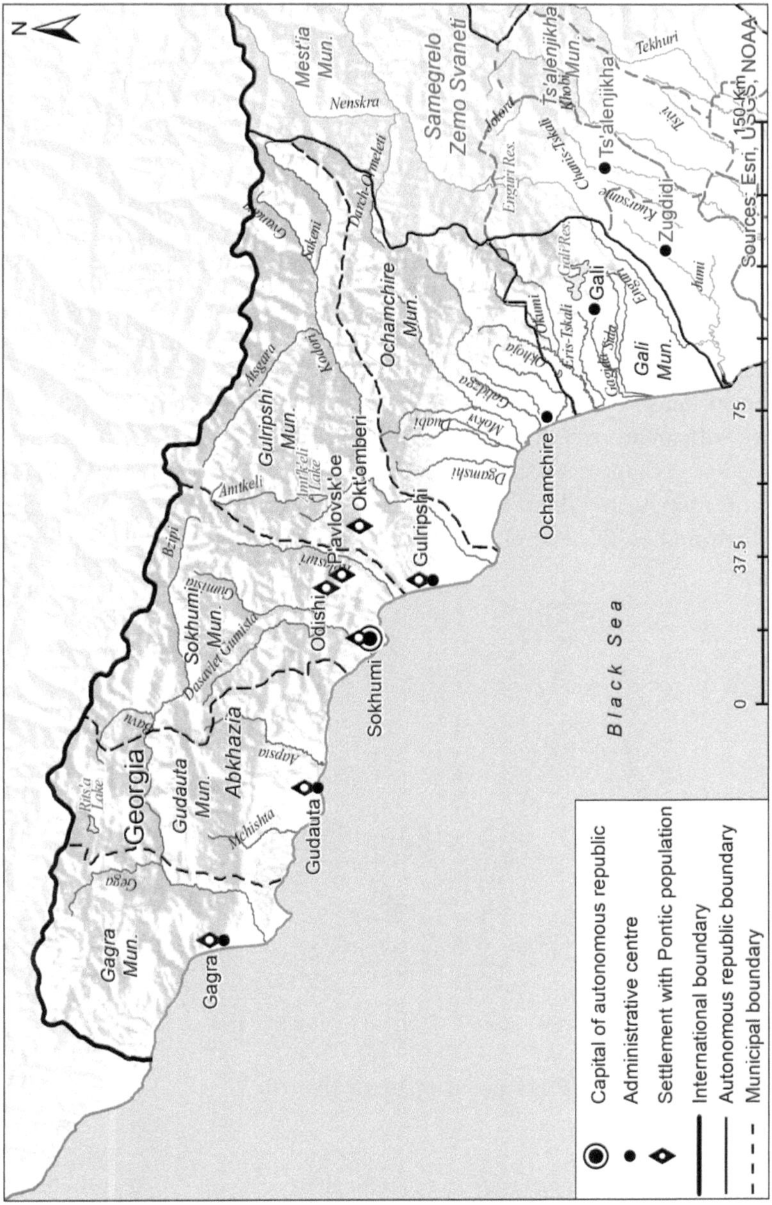

Map 5 (Former) Greek compact settlements in Abkhazia

period. The number of ethnic Georgians also grew, due to the resettlement of Megrelians and Svans (see Table 4).

By the end of the 19th century, the aforementioned migration processes had resulted in the ethnic Greek population of the Abkhazian territory exceeding 5,000 inhabitants, making up 5.08% of the total population of the region. Afterwards, the number of Pontic Greeks in Abkhazia grew steadily, mainly due to immigration. This reached a peak in 1939, with 34,621 inhabitants or 11.1% of the region's total population (see Table 4, Figure 7).

The 1939 data are the last to illustrate the mechanical growth of the Greek population, as emigration of the Greek population from the newly-founded Republic of Turkey had ceased in the end of the 1920s. The termination of this immigration process in Georgia can be attributed to two main factors. First, the forced Sovietization of Georgia played an important role, as a result of which its international land and sea borders were gradually closed.[20] The second factor is the conclusion of the Greco-Turkish conflict. Ultimately, the 1919–22 conflict ended with an unsuccessful Greek intervention and the 1923–24 agreement on population exchange. As a result of this agreement, ethnic Turks living in Greece, and millions of ethnic Greeks living on Turkish territory, were forced to leave their homes. Because of its scale, this great resettlement was supposed

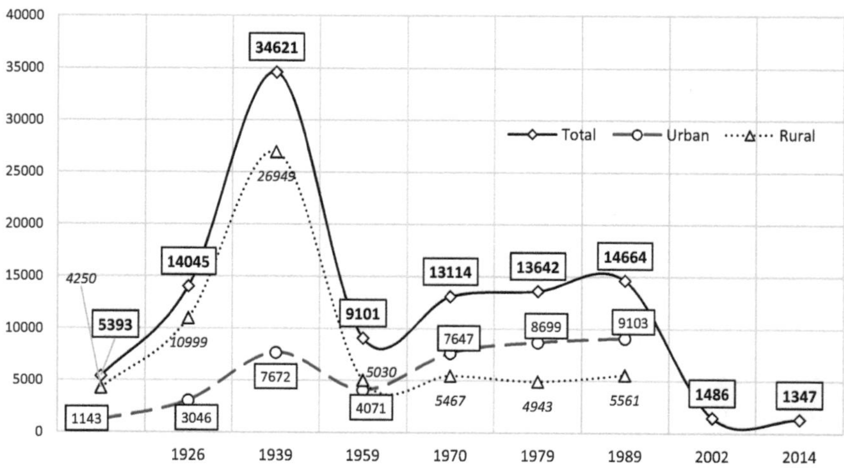

Figure 7 Dynamics of Greek population in Abkhazia 1897–2011

20 During the Soviet occupation of Georgia, only the land border with Turkey was closed, whereas bordering Russia, Armenia, and Azerbaijan were part of the Soviet Union.

Table 4 Ethnic composition of the population in Abkhazia, 1897–2011

Year	Total	Abkhaz.	%	Georgian	%	Armen.	%	Russian	%	Greek	%	Ukrain.	%	Other	%
1897[1]	106 179	58 697	*55.2*	25 873	*24.3*	6 552	*6.1*	5 135	*4.8*	5 393	*5.0*	809	*0.7*	3 720	*3.5*
1926[2]	201 016	55 918	*27.8*	67 494	*33.5*	25 677	*12.7*	12 553	*6.2*	14 045	*6.9*	4 647	*2.3*	20 680	*10.2*
1939[3]	311 885	56 197	*18.0*	91 967	*29.4*	49 705	*15.9*	60 201	*19.3*	34 621	*11.1*	8 593	*2.7*	10 601	*3.4*
1954[4]	404 738	61 193	*15.1*	158 221	*39.0*	64 425	*15.9*	86 715	*21.4*	9 101	*2.2*	11 474	*2.8*	13 609	*3.3*
1970[5]	486 959	77 276	*15.8*	199 596	*40.9*	74 850	*15.3*	92 889	*19.0*	13 114	*2.6*	11 955	*2.4*	17 279	*3.5*
1979[6]	486 082	83 087	*17.0*	213 322	*43.8*	73 350	*15.0*	79 730	*16.4*	13 642	*2.8*	10 257	*2.1*	12 694	*2.6*
1989[7]	525 061	93 267	*17.7*	239 872	*45.6*	76 541	*14.5*	74 913	*14.2*	14 664	*2.7*	11 655	*2.2*	14 149	*2.6*
2003[8]	214 016	94 597	*44.2*	44 041	*20.5*	44 869	*20.9*	23 420	*10.9*	1 486	*0.6*	1 797	*0.8*	3 806	*1.7*
2014[9]	242 756	123 767	*50.9*	46 645	*19.2*	41 944	*17.2*	22 280	*9.1*	1 347	*0.5*	1 759	*0.7*	5 014	*2.0*

1 Source: Troynitsky, Nikolay (1905) *Pervaja vseobŝaja perepis' naselenija Rossijskoj Imperii 1897g. (The first universal census of the population of the Russian Empire in 1897. Translation NL)* Vol. LXVI. Saint Petersburg: Ripol, 88-93 (In Russian).

2 All-Union Population Census (1926b) Nacional'nyj sostav naselenija po regionam respublik SSSR (National composition of the population by regions of the republics of the USSR. Translation NL). Retrieved from http://www.demoscope.ru/weekly/ssp/sng_nac_26.php?reg=2346 (in Russian).

3 Source: All-Union Population Census (1939b) *Nacional'nyj sostav naselenija rajonov, gorodov i krupnyh sel sojuznyh respublik SSSR* (National composition of the population of districts, cities and large villages of the union republics of the USSR. Translation NL). Retrieved from http://www.demoscope.ru/weekly/ssp/ussr_nac_39_ra.php?reg=823 (in Russian).

4 Source: All-Union Population Census (1959b) *Gorodskoe i sel'skoe naselenie oblastej respublik SSSR (krome RSFSR) po polu i nacional'nosti* (Urban and rural population of regions of the republics of the USSR (except for the RSFSR) by gender and nationality. Translation NL). Retrieved from http://www.demoscope.ru/weekly/ssp/resp_nac_59.php?reg=62 (In Russian).

5 Source: All-Union Population Census (1970b) *Gorodskoe i sel'skoe naselenie oblastej respublik SSSR (krome RSFSR) po polu i nacional'nosti* (Urban and rural population of regions of the republics of the USSR (except for the RSFSR) by gender and nationality. Translation NL). Retrieved from http://www.demoscope.ru/weekly/ssp/resp_nac_70.php?reg=68 (In Russian).

6 All-Union Population Census (1979b) *Gorodskoe i sel'skoe naselenie oblastej respublik SSSR (krome RSFSR) po polu i nacional'nosti* (Urban and rural population of regions of the republics of the USSR (except for the RSFSR) by gender and nationality. Translation NL). Retrieved from http://www.demoscope.ru/weekly/ssp/resp_nac_79.php?reg=71 (In Russian).

7 All-Union Population Census (1989b) *Gorodskoe i sel'skoe naselenie oblastej respublik SSSR (krome RSFSR) po polu i nacional'nosti* (Urban and rural population of regions of the republics of the USSR (except for the RSFSR) by gender and nationality. Translation NL). Retrieved from http://www.demoscope.ru/weekly/ssp/resp_nac_89.php?reg=65 (In Russian).

8 Source: UGSRA (2019) *Abhazija v Cifrah za 2014 god* (Abkhazia in numbers for 2014. Translation NL). URL: https://ugsra.org/abkhaziya-v-tsifrakh/2014-god.php (In Russian).

9 Ibid.

to definitively resolve the ethnic tensions found in these countries by maximally reducing the populations of specific ethnicities.

As discussed above, immigration was the main cause of the high growth rate of the Greek population, and the compact settlements were mainly located in rural areas. According to the 1939 census, the number of Greeks in Abkhazia reached a historic high with a population of 34.621, corresponding to 11% of the region's total population. 1939 data also show the largest recorded difference between urban and rural populations. As seen in Figure 7, in 1939 77% of the Greek population in Abkhazia (26,949 individuals) were living in villages and 23% (7,672) in cities.

The 1939 and 1959 censuses are particularly interesting, since they illustrate the first negative effect of mechanical migration. The first radical decline of the Greek population in Abkhazia took place during this period, when their numbers declined by 73.7% (25,520 individuals). This was a result of the repressive Soviet policies carried out on social and ethnic grounds in the 1940s. Deportations were especially harsh for compact, rural Greek settlements, whose population declined by 81%, from 26,949 to 5,030 individuals. The decline of the urban Greek population was 47%, from 7,672 to 4,071 individuals.

The next important change for the Greek community in Abkhazia came, as above, with the fall of the Soviet Union and the ensuing social, economic, and political crises and military conflicts. As a consequence, according to the *de facto* State Statistics Service of Abkhazia, the number of Pontic Greeks in the region declined by 90% between 1989–2011 (see Figure 7, Table 4).

3.6 Adjara

The migration of Anatolian Greeks to this region of western Georgia, on the coast of the Black Sea, is also related to the Russian Empire's ethnic policies. It is generally accepted that an ethnic Greek community had already existed in Adjara, and specifically in the port city of Batumi, since the first half of the 19th century. However, the Russian Empire organized further migrations between the 1880s and 1920s, resettling Greeks on lands forcibly vacated through Muhajirism (Garakanidze 2000; Mavrogordato 1931).

The oldest documents attesting to a Greek population in of Batumi date to 1835. During that period, Adjara was part of the Ottoman Empire, and according to the "Census of the male population from the Churuksu (Kobuleti) Sanjak, part of the Childiri Vilayet", a document created in November 1835 and kept at the Prime Minister's Ottoman Archives, temporarily-settled Greeks also worked as merchants and goldsmiths alongside Georgians at Kobuleti's market (Giorgadze et al. 2017:21–22).

The Neo-byzantine Orthodox Church of Saint Nicholas also attests to the presence of a Greek community in the city. Still standing today, Batumian Greeks received authorization for the church's construction in 1865, under the rule of the Ottoman Empire. The church is located in the "Urum neighborhood" (*Urum Mehele*), which was partly inhabited by Greeks, and was probably the first compact settlement of the Greek community in Adjara. This fact disproves the idea, forwarded by Giorgadze et al. (2017), that there were no compact Greek settlements on Adjarian territory until 1878.

As in Abkhazia, the wave of immigration into Adjara which began in 1877–1878 was the direct consequence of the military conflict between the Russian and Ottoman empires. The conflict stopped temporarily in January 1878 through the ceasefire agreement of Adrianople, and then ended in March of the same year with the Treaty of San Stefano. According to the Treaty, the victorious Russian Empire would incorporate Ardahan, Kars, Doğubayazıt (formerly Bayazit) and Adjara, including Batumi (Kokoev et al. 1999; Höfler 2017; Gachechiladze 2008).

The change of borders between the two empires was followed by a new ethnic policy in Adjara, with a Muhajirism process similar to that which took place Abkhazia. Although some of the population that emigrated to the Ottoman Empire could not assimilate to the new environment and living conditions, and therefore decided to return, it is commonly accepted that as a result of the 1878 Muhajirism, Adjara's

Muslim population declined by about 8,000 individuals. As in Abkhazia, the Russian Empire encouraged the immigration of Christian Greeks from the Ottoman Empire into the land left vacant (Jaoshvili 1968:113; 1984:90).

The majority of the Greek immigrants were Pontic, and therefore hellenophone. They mainly came from Trabzon, Rize, Pakha, Ordu, Giresun, and Samsun. Some must have been from the Santa community (Greek Σάντα – Santha/Santa, Turkish Dumanlı) in the Gümüşhane Province, given the naming of the village Santa founded in the Ts'alk'a Municipality in Kvemo Kartli (see Map 3) (Pashaeva 1992:2; Giorgadze et al. 2017).

Pontic Greeks established their first settlements in Adjara after 1881, near Batumi and Kobuleti. According to several sources, one of the first Greek compact settlements in Adjara was the village of Dagva. Later, with the increase in Greek immigration, they also established compact settlements in other villages (five in total). With Adjara's lowlands experiencing a malaria epidemic at the time, Greeks chose to establish their compact settlements at some distance from the shore, on territories about 200–400 meters above sea level (see Map 6, Annex 1) (Garakanidze 2000; Giorgadze et al. 2017).

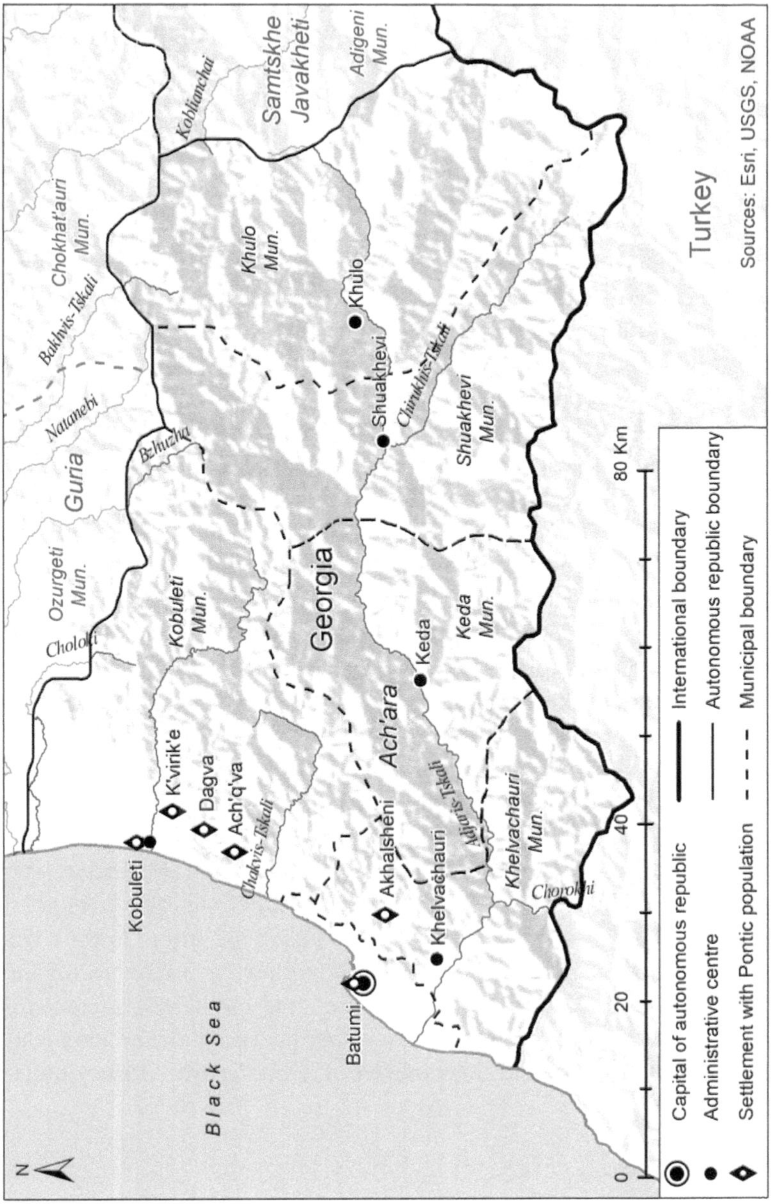

Map 6 (Former) Greek compact settlements in Adjara

When discussing the reasons behind Greek immigration to Adjara, one must also consider the rapid urbanization and economic development processes that were taking place at the end of the 19th century and the beginning of the 20th century, in Batumi as well as more generally in the South Caucasus. In the case of Batumi, the growth of the city was linked to the development of transportation infrastructure, specifically the deep-water port. Moreover, in 1883 Batumi was connected by railway to Baku, which was the main industrial hub of the Caucasus at the time. The railway also passed through Tbilisi. Later, between 1897–1907, the world's first oil pipeline was constructed between Baku and Batumi. This lent Batumi's port particular strategic and economic importance, further accelerating industrial and commercial development. This, in turn, led to a major increase in labor immigration from bordering regions, which involved Pontic Greeks along with other ethnic groups (Jaoshvili, 1984; Gachechiladze, 2008).

As in the rest of Georgia, the Greek population in Adjara saw steady growth between 1897–1989, although its share in the total population steadily declined. Accordingly, during this period the number of Greeks in Adjara only increased from 4,650 (5.4% of the total population) to 7,396 (1.8%), its share in the population having peaked at 7.0% in 1939 (see Table 5, Figure 8).

Several particular circumstances apply to Greek population dynamics in western Georgia. As noted above, from the second half of the 19th century until the end of the 1920s, population growth stemmed mainly from immigration, or in other words, mechanical movement. During the same period, in the context of the ongoing Russo-Turkish and Greco-Turkish wars, which were leading to the forced relocation of some populations on ethnic and religious grounds, the evacuation of Greek individuals who had Greek nationality[21] was also organized from the port of Batumi. This was with the support of the Greek government and the Greek diaspora (Giorgadze et al. 2017). Because of its geographic location, Batumi's port was one of the most important destinations for Greeks who had emigrated from the Ottoman Empire's eastern *vilayet* and who were often refugees. From Batumi, some headed to Greece, and others moved to the territories of modern Russia and Ukraine. Another important part stayed in Adjara. The abovementioned processes were behind the rapid mechanical increase of the Greek population in Adjara, which ceased after the end of the military conflicts on Anatolian territory and the closure of the border with Turkey under Soviet rule.

21 Note: Ethnic Greeks who hold the Greek nationality will be discussed in Chapter IV.

Table 5 Ethnic composition of the population in Adjara, 1897–2014

Year	**Total**	**Georgian**	%	**Russian**	%	**Armenian**	%	**Greek**	%	**Ukrainian**	%	**Abkhazian**	%	**Other**	%
1897	88 444	57491	*65.0*	7 217	*8.1*	7 120	*8.0*	4650	*5.2*	1 637	*1.8*	687	*0.78*	9 642	*10.9*
1926	131 957	90314	*68.4*	10 190	*7.7*	5 343	*4.0*	5 547	*4.0*	1 500	*1.1*	779	*0.59*	18 284	*13.86*
1939	200 106	127542	*63.7*	30 535	*15.2*	1 4085	*7.0*	7 959	*3.9*	6 866	*3.4*	1 029	*0.51*	12 090	*6.04*
1959	245 286	178 661	*72.8*	32 794	*13.3*	15 830	*6.4*	5 737	*2.3*	5 844	*2.3*	1 157	*0.47*	5 263	*2.15*
1970	309 768	236 928	*76.4*	35 774	*11.5*	15 614	*5.0*	6 867	*2.2*	7 181	*2.3*	1 361	*0.44*	6 043	*1.95*
1979	354 224	283 872	*80.1*	34 544	*9.7*	16 101	*4.5*	7 072	*2.0*	5 402	*1.5*	1 508	*0.43*	5 725	*1.62*
1989	392 432	324 813	*82.7*	30 042	*7.6*	15 849	*4.0*	7 396	*1.8*	5 943	*1.5*	1 636	*0.42*	6 753	*1.72*
2002	376 016	351 132	*93.3*	9 073	*2.4*	8 848	*2.3*	2 168	*0.5*	1 056	*0.2*	1 558	*0.41*	2 181	*0.58*
2014	333 953	320 742	*96.0*	3 679	*1.1*	5 461	*1.6*	575	*0.1*	793	*0.2*	297	*0.09*	2 046	*0.61*

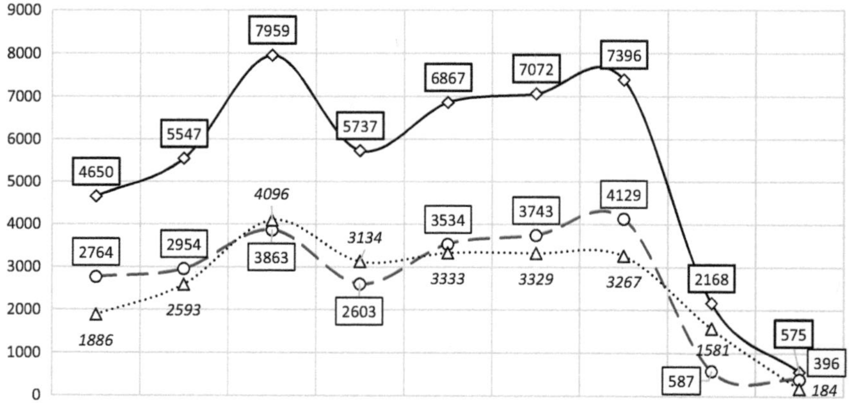

Figure 8 Dynamics of Greek population in Adjara for 1897–2014 according to the data used in Table 5

As noted in subchapters 3.2 and 3.5, the sharp decline of Greek population numbers in both Adjara and Abkhazia (as well as other Soviet Black Sea regions) is linked to the Soviet repression and deportations of Greeks to Central Asian republics in 1937–1952. These are discussed in more detail in the next chapter.

From the 1950s until the collapse of the Soviet Union, Adjara's Greek population saw steady natural growth, but a proportional decrease due to the immigration of Russian, Armenian, Ukrainian and other nationalities to Adjara during 1926–1957 and the region's overall high birth rate.

Apart from the decline linked to deportations, Adjara's Greek population grew slowly but steadily until 1989. This radically changed with the collapse of the Soviet Union, and from 7,396 inhabitants in 1989 their numbers decreased to 2,168 in 2002 (0.55% of the total population) and to 575 (0.1%) in 2014. The reason behind this 92% drop is the mass emigration discussed in Chapter 5.

This chapter has presented three important arguments. First, that the immigration of Greeks to Georgia was mainly a result of agreements between states, which also influenced the specificities of their settlements. Second, that the most notable difference between Pontic and Urum Greeks is that they were settled in different regions: Pontic Greeks on the Black Sea coast in western Georgia, and Urum (with a relatively small number of Pontic Greeks) in eastern Georgia, in Kvemo Kartli and Samtskhe-Javakheti. Third, that this spatial distribution played an important role during the 1937–53 repression and deportations, which affected only the Pontic population on the Black Sea.

Chapter IV. The Greek Community of Georgia and Soviet repressions

In Georgia, as in other Soviet republics, political repression began as soon as the Bolshevik regime was firmly established and continued, with varying intensity, until the collapse of the Soviet Union. It included the mass deportations of the 1940s, which significantly influenced Georgia's demographic, ethnic, and social structure, and thus the fate of Georgian Greeks.

The history of these acts of repression is complex and requires careful reconstruction, well beyond the scope of the present work. Instead, this chapter focuses on the ethnic components of Soviet repression, whereby forced migration led, according to the censuses of 1939 and 1957, to the first mechanical decline in the Greek population in Georgia. This took place specifically in Adjara and Abkhazia (see Figures 1, 7, 8). At the same time, these tragic demographic processes were accompanied by the closure of Greek institutions, schools, and newspapers, which also left its mark on the Greek community's potentialities – demographic as well as social and cultural (Pohl 1999: 123; Gachechiladze 2008; Popov 2016).

First, I briefly discuss the various stages of Soviet repression and their relevant aspects, and provide an overview of related literature. In the subchapters that follow, I trace the historical background of the acts of repression that directly affected the Greeks of Georgia: the *dekulakization* in 1928–33 (4.1), the *National Operations* of 1937–38 (4.2), deportations during the 1940s and 1950s (4.3), and destalinization, i.e. the *Khrushchev Thaw* (4.4). I describe these stages' influence on the dynamics of the Greek population's size in Georgia and on the specific features of their deportations (4.3). Finally, I analyze contemporary perspectives on Soviet repression in light of our interviews (4.4).

During the first stages of the country's Sovietization (1921–1928), persecutions were for the most part socially and politically motivated. Repression affected all levels of society – the opposition, the government of the Democratic Republic of Georgia, former government officials and military personnel, ideologically non-conformist intelligentsia, and the nobility and farmers who took part in the 1924 August Uprising. Despite a brief period of independence (1918–1921), Georgia was also affected by the *Red Terror* that swept Russia during the Civil War. After the establishment of the Bolshevik regime, the Red Terror started throughout the Caucasus – even in countries that were latecomers to the Soviet Union. It is thought that several thousand people fell victim to the Terror, and that up to 20,000 people were sent into exile during the first years of Sovietization (Gachechiladze 2008; pp 242–3).

The next stage of repression was the collectivization reform conducted in 1928–33, also known as "*dekulakization*". The declared objective of the program was to get rid of prosperous peasants as a class, and to carry out the expropriation of privately-owned agricultural property. Acts of repression became particularly ruthless in the second half of the 1930s. The main goal of the Stalinist regime was to punish segments of society which it found politically or ideologically unacceptable. The terror was most intense in 1937–38. The first wave of repression against specific ethnic groups, known as "*National Operations*", also took place during these years. Their conclusion was officially announced in 1938, even though persecution on ethnic grounds continued into the 1940s, when it took on the form of mass deportations.

The population transfer that started in the 1940s in the USSR was mainly linked to preparations for World War II and continued during the war. Mass deportations, and mass repression more broadly, persisted until 1953 and only ended after Stalin's death (Rummel 2017; Junge, Bonwetsch, Müller 2015).

From 1953 to the beginning of the 1960s, the new Soviet leader Nikita Khrushchev promulgated a policy of destalinization (also known as the "*Khrushchev Thaw*" or *Ottepel*). During that time, repression and censorship were less common, and millions of political prisoners and sanctioned ethnic groups were "rehabilitated". This resulted in an important decline in the incarcerated population of the gulags and "special institutions". Nonetheless, during the rehabilitation of *National Operations* victims, ethnicity remained a determinant social factor, since the deportees were usually not given the opportunity to return to their own homes or home regions (Pohl 1999; Gachechiladze 2008). Despite destalinization, repression still took place on a smaller scale in the USSR during this period and further through Brezhnev's *Stagnation* and Gorbachev's *Perestroika*.

The Gulag system was officially terminated in the 1960s, but the deportation of political prisoners to labor correction colonies maintained the same type of penitentiary infrastructure and its principles. While it took place on a much smaller scale following Stalin's rule, severe political persecution continued, its intensity often depending on the USSR's economic, political and international situation. However, unlike the 1930–50s, mass repression on ethnic grounds no longer took place.

Information about the repression of the 1930–50s was extremely difficult to acquire until the 1980s, and only existed in the memories of direct victims and their families. The subject was taboo in Soviet academia and in society as a whole until *Glasnost*[22] in the 1980s. Although over 30 years have now passed since *Glasnost* as well as the end of the Soviet Union, academic study of Georgia's Soviet past is far from comprehensive, with research on ethno-national repression particularly lacking.

Instead, studies so far have focused on the *Great Terror* and in particular on the tragic fate of the Georgian elite, its intellectuals and famous public figures (Junge, Tushishvili, Bonwetsch 2015). The two recent exceptions are "Bolshevik Order in Georgia. The Great Terror in a small Caucasian republic," (Junge, Tushishvili, Bonwetsch 2015) and "Ethnos and Terror in Soviet Georgia" (Junge, Bonwetsch, Müller 2015). These studies represent an attempt to analyze the mass repression of the 1930s from a different perspective, concentrating on those carried out against ethnic minorities, their main features and possible motives.

While only the compactly settled Greek population that lived in the Black Sea region was affected by deportations, they were not carried out in isolation from the *National Operations* taking place throughout the Soviet Union. Hence, I cover the operations conducted throughout the Soviet Union, the Georgian SSR, and the Black Sea region that are linked to the Greek population.

4.1 1928–33 Dekulakization

Repressions (political, social, and ethnic) in Georgia and the Soviet Union took place throughout the Soviet Union's existence (Rummel 2017). Punitive operations conducted against specific ethnic groups started in 1937–38, and the dekulakization in Georgia was not overtly ethnically motivated.[23] Nonetheless,

---

22 Glasnost, (Russian: "openness") was a Soviet policy of open discussion of political and social issues. It was instituted by Mikhail Gorbachev in the late 1980s and began the democratization of the Soviet Union. Ultimately, it caused fundamental changes in the political structure, reducing the power of the Communist Party and leading to multi-candidate elections. Glasnost also permitted criticism of government officials and allowed the media more freedom to disseminate news and information .

23 By 1932–33, collectivization had led to famine, particularly affecting Ukraine and Kazakhstan. Starvation and disease killed between six and ten million people in Ukraine alone, where the period is known as the Holodomor (Ukrainian: Голодомóр "to kill by starvation"). A massive famine also struck the South Urals, the Lower Volga Region, the modern-day Rostov Oblast, and Kazakhstan, where according to some accounts it wiped out almost 40% of the Kazakh gene pool (Gachechiladze 2008: 290–1). Many

dekulakization remains an important component in the study of the repression suffered by the Greek population of the Soviet Union and the Georgian SSR specifically.

As mentioned, the wave of collectivization/dekulakization in 1928–33 aimed to eliminate prosperous peasants and private agricultural property. A second objective was to colonize resource-rich territories in Siberia, Kazakhstan, the Urals, and Northern Russia at the expense of the repressed populations. This plan was openly announced in 1929, in the first five-year plan of the USSR (Rummel 2017; Gachechiladze 2008).

The main phase of dekulakization did not discriminate on an ethnic basis. Rather, it used ideological criteria. Therefore, it equally affected all citizens of the Georgian SSR, including Greeks. However, Greeks holding foreign citizenship and living in the Soviet Union did not have to go through the first and most severe stage of collectivization. In addition, the scale of dekulakization in the Caucasus was small compared to the process in Eastern Ukraine, Southern Russia, and Kazakhstan. This can be attributed to several factors:

- A large share of the Greeks in the Soviet Union retained their foreign (mostly Greek and then Turkish) citizenship, and refused to acquire Soviet citizenship. At this stage of political repression, the Soviet Union did not dare to openly persecute foreign citizens, given its still-weak international status (Dzhukha 2006);
- While collectivization was ruthless in every Soviet republic, the most severe repression took place in agricultural areas that focused on grain production (Eastern Ukraine, modern-day Rostov Oblast, Kuban).[24] In the Caucasus, the soil and slope of the land makes it unfit for grains, hence escaping this circumstance (Gachechiladze 2008:290–1);[25]
- The majority of Georgia's Greek population were impoverished first or second-generation refugees, mainly living in rural settlements. Therefore, few

---

scholars unequivocally consider the Holodomor an act of genocide, with both class and racial characteristics (Courtois et al. 1999:159–61; Rummel 2017: 95–96; Mace 1986:11).

24 In the case of Kazakhstan, mass famine was caused by the confiscation of the nomad population's livestock and their forcible settlement (Gachechiladze 2008; p 290–1).

25 However, the same republics were ordered to grow other labor-intensive crops. For instance, in order to attain "tea independence", the subtropical regions of Western Georgia had to cultivate tea, while Azerbaijan and Central Asian republics had to develop cotton-growing in order to reach "cotton independence" (ibid).

if any Greeks had amassed enough property to be considered "kulaks" (Junge & Müller 2017).

In 1937–38, however, ethnicity became involved as an explicit motive for repression. During the next, instead of saving them, Greek individuals' foreign citizenship became a reason for repression and a "capital offense". Dekulakization was also still used as a rationale, but on a smaller scale.

4.2 1937–38 NKVD's Greek Operation

The culmination of the Soviet *Great Terror* took place in 1937–38, targeting "former kulaks, criminals, and other anti-Soviet elements". Based on NKVD Order N° 00447, it soon took on ethnic features. Some scholars associate this with the Stalinist transformation of the Soviet regime's political paradigm, whose basis shifted from social class to ethnicity (Savin 2017; Junge et al. 2015).

During the NKVD's 1937–38 *National Operations*, numerous minorities were persecuted including Poles, Germans, Finns, Estonians, Latvians, Romanians, Bulgarians, Macedonians, Greeks, Iranians, Afghans, Chinese, and Koreans. According to the latest data, these operations' death toll was between 340,000 and 350,000 people, mostly Poles (about 140 000) and Germans (about 55 000) (Savin 2017: 48; Pohl 2000:23). The former was the first group to be accused, as a whole, of sabotage and espionage on behalf of the Polish government, in Order N° 00485 signed by Nikolai Yezhov, the USSR's People's Commissar for Internal Affairs, on August 9, 1937. Later, such accusations would also be put forward against people of German, Latvian, and Greek origin, among others (Savin 2017; Dzhukha 2006).

The NKVD's *National Operation* against Greeks was initiated with Order N° 50215, signed by Yezhov on 11 December, 1937. The document was declassified by the Central Archive of the FSB on April 12, 2006 thanks to the efforts of scholars Ivan Dzhukha and Ivan Savvidis, who quote as follows (Dzhukha 2006; 51–52):

> [...] Subject to arrest are all Greeks (Greek nationals and citizens of the USSR) of the following categories: a) Who are on the operational record; b) Former major traders, speculators, smugglers and currency dealers; c) Greeks leading active anti-Soviet, nationalistic activity, primarily those who have already been dekulakized or are hiding from dekulakization; d) Political migrants from Greece and all Greeks who illegally arrived in the USSR, regardless of the country of origin; e) All Soviet Greeks who are so-called foreign agents of the NKVD and the intelligence department of the Red Army.

As with other ethnic groups, the accusations in this period no longer stemmed from individual victims' past, "inappropriate" ideological views or lawbreaking; the main factor was ethnic belonging, with the order being directed at the Greek minority in general. Unlike the former stages of repression, holding foreign citizenship now made people targets rather than keeping them safe.

The first and only attempt to analyze data about the numbers and ethnic background of victims of the *National Operations* in Georgia was carried out in collaboration between two academic research projects, *"Stalinism in the Soviet Provinces 1937–1938"* and *"Mass Repression in Soviet Georgia"*, using the archive of the Georgian Ministry of Internal Affairs (see Table 6). This data provides a more accurate picture of the repression's ethnic basis and allows a comparison of the scale on which different ethnic groups were targeted.

As can be seen, Germans suffered the largest proportional losses among ethnic minorities repressed in Georgia (2.1% of the Germans living in Georgia), followed by Abkhazians (0.99%), and Jews and Greeks (both 0.9%). The relative share of repressed ethnic Georgians is lower (0.67%).

Contextually, the Soviet Union was already preparing for war against Nazi Germany, hence the intensive targeting of German minorities – especially in coastal regions – which took the form of mass deportations during the 1940s.

Compared to the German case, the repression of Abkhazians requires more complex analysis. In this case, Junge and Müller rely only on quantitative data from 1937–38 and highlight the ethnic policy without considering the historical context. Rather, they assume that the Georgian SSR government took advantage of the *National Operations* initiated by Moscow in order to repress Abkhazia's non-Georgian populations (including Greeks) and thereby extend Tbilisi's power in the region (Junge & Müller 2017; Junge et al. 2017). As mentioned above, this idea offers a completely new and interesting perspective, yet the quantitative analysis does not take into account the spatial and chronological conditions of the repressions, two factors which greatly influence the data.

When comparing the repressed persons according to ethnic belonging, certain events and the context of the period in question have to be taken into account. For instance, the high relative number of Germans was due to the fact that the Soviet Union was preparing for war against Nazi Germany. Therefore, Stalin's regime considered the German population of the coastal regions as particularly dangerous. Their persecution, just like that of the rest of "untrustworthy nationalities", became mass deportations during the 1940s. These once again mainly affected the ethnic minorities living in coastal regions.

Compared to the German case, the repressions of Abkhazians require a relatively more complex analysis. In this case, Junge and Müller only rely on the

Table 6 Degree of repression during all mass operations 1937–1938 (kulak troika[1], dvoika[2], national troika, police troika[3]) (Source: Junge & Müller, 2016 pp 125)

Nationality	Population	Repressed persons 1937–38		Population/Repressed persons % 1937–1938	
		Total	Executed	Total	Executed
Titular Nationality					
Georgians	2 080 179	13 986	6 169	0.67	0.30
Lesser titular nationalities					
Abkhazians	55 409	548	308	0.99	0.55
Ossetians	143 604	384	207	0.27	0.14
Diaspora nationalities					
Greeks	87 385	798	168	0.91	0.19
Turks	82 500	520	231	0.63	0.68
Jews	29 721	264	124	0.9	0.43
Germans	24 135	514	225	2.1	0.94

1 "***Kulatskaya troyka*** and ***natsional'naya troyka***: An extrajudicial committee of three, formed in the capital and in all large administrative units of the USSR, presided over by the respective head of the secret service, the NKVD. Members were the corresponding prosecutors and Party secretaries." (Junge & Müller, 2016 pp 125)

2 "A ***dvoika*** was an extrajudicial commission of two, which was composed of the federal, regional or local state prosecutor and the current head of the NKVD" (Ibid).

3 "***Militseyskaya troyka*** ['police troika']: The chairman was usually the local, regional or federal head of the NKVD or his representative, the other members being the corresponding state prosecutor and the leader of the administration of the civil police (Ru. militsiya), as well as the leader of the responsible departments of the police (Ibid)".

quantitative data from 1937–38, and only highlight the ethnic policy. They do so without considering the historical context. Rather, they assume that the government took advantage of the *National Operations* Moscow initiated and encouraged repressions against the non-Georgian population (including Greeks) of Abkhazia in order to spread their central power (that of Tbilisi) to the region (in Abkhazia) (Junge & Müller 2017; Junge et al. 2017). As I have mentioned, this idea offers a completely new and interesting perspective, even though the quantitative analysis does not take into account the spatial and chronological conditions of the repressions, two factors which greatly influence the quantitative data.

When discussing the repressions in Abkhazia during 1937–38, the specificities of the dekulakization process are important. Dekulakization/collectivization started much later in Abkhazia compared to the rest of Georgia. It started in 1936, and in some areas, lasted until 1938. The rest of the Georgian SSR went through the hardest phase of dekulakization during 1929–1932 (Derluguian 1998). These circumstances are important, as according to the data presented in the table, all the repressed persons in Abkhazia (including 178 Greeks) were tried by a kulak *troika* (Junge & Müller 2017:128).

When discussing repression in Abkhazia during 1937–38, the specifics of the dekulakization process are important. Dekulakization/collectivization started much later in Abkhazia compared to the rest of Georgia – beginning in 1936, and in some areas lasting 1938. The rest of the Georgian SSR had already gone through the harshest phase of dekulakization between 1929–1932 (Derluguian 1998). These circumstances are important, since according to the data presented in the table, all the repressed persons in Abkhazia (including 178 Greeks) were tried by a kulak *troika* (Junge & Müller 2017:128).

The *National Operations* ended in November 1938, after Lavrenty Beria, previously First Secretary of the Georgian Communist Party, replaced Nikolai Yezhov as head of the NKVD. Notably, Yezhov himself did not escape Stalin's repressions, and was arrested on conspiracy charges in 1939 and executed the following year.

In the end, determining the exact number of Greek *National Operations* victims is very difficult and requires additional research, as current scholarly estimates still range widely: between 800–3,700 in Georgia and between 6,500–22,000 in the Soviet Union as a whole (Junge & Müller 2017:129; Courtois et al. 1999:147; Danilov 2011:721).

4.3 Deportations

Despite the end of the *Yezhovshchina* (the period of intense repression under Yezhov), Stalin's regime did not put an end to ethnic persecution. On the contrary, during 1939–1953 it took it to a new level through mass deportations.

The deportations were directly linked to the Second World War, and their causes are conventionally divided into "preventative" deportations of individuals of "untrustworthy" nationality and" punitive" deportations of entire ethnic groups which were "tried" and declared collectively guilty of collaboration with Nazi Germany. In both cases, the strategic goal was the same: to ethnically homogenize and/or create a loyal ethnopolitical balance in border territories (Gachechiladze 2008).

During these forced mass migrations, groups subject to total deportation included Koreans, Finns, Germans, Kalmyks, Karachays, Chechens, Ingush, Balkars, Crimean Tatars, Greeks, Kurds, Meskhetian Muslims, and Hemshin (Polian 2003; Gachechiladze 2008; Junge et al. 2015; Popov 2015). Ethnic groups were deported from border territories to unexploited regions of the inhospitable Soviet hinterland – Kazakhstan, Kyrgyzstan, and the Urals (See Map 7).

In the case of Soviet Georgia, the main victims of mass deportation were Germans. On the 8th of September 1941, ten weeks into the Nazi invasion, a directive was issued for the deportation of 48,609 Germans living in Transcaucasia, due for completion by 15–30t of October. During the following years, deportations also affected other nationalities, particularly on the border territories of Soviet Georgia (Gachechiladze 2008; Polian 2014; Jishkariani 2018).

Deportations of Greeks took place throughout the Soviet Union, in three main stages: during 1942, 1944, and 1949. First, Stalin ordered the deportation of over 1,500 Pontic Greeks and other minorities from the Krai of Krasnodar in May 1942. Then, in 1944, as a result of direct pressure from Beria's deputies Ivan Serov and Grigol Karanadze, over 15,000 Soviet Greeks were deported from Crimea and another 8,300 from the Krasnodar Krai and the Rostov region. A further 16,375 Greeks were relocated from Georgia, Armenia, and Azerbaijan to Kazakhstan and Russia (Polian 2001:116; Popov 2016: 60–61).

Finally, in 1949 the Soviet Council of Ministers relocated the remaining Greeks, as well as Muslims of various ethnic backgrounds, from the Black Sea coast and the Southern Caucasus (mainly Georgia and Armenia) to Kazakhstan. The estimated number of Greeks among these deported groups vary between 27,000 and 36,000. The repression of 1949 notably affected former Greek citizens who had already become Soviet citizens (Bugay 1996:92; Kotsonis 1997:87).

Unlike other nationalities (Crimean Tatars, Ingush, Chechens, Kalmyks), Greeks were not accused of collaborating with Nazi Germany. Stalin's regime targeted Greeks as "untrustworthy" based on a number of factors: the large number of Greeks who (had) held foreign passports (mostly Greek and Turkish/Ottoman); their settlements' location in coastal regions or port cities; and their foreign family and social connections (See Pictures N° 1, 4, 5). For instance, 1926 census data which recorded foreign nationality suggest there were 22,287 foreign citizens living in Georgia that year, most of whom were Greek. Of these, most (82%) were registered in Abkhazia (3,284) and Adjara (15,012). It thus appears that the deportations were a *preventive* measure – paranoid as it was – driven by social, political, and geographic factors.

This idea is supported by the fact that mass deportations were only conducted in coastal regions, and did not affect, for instance, the Urum population of Kvemo Kartli (See Map 8). Unlike the coastal regions of the Black Sea, Kvemo Kartli was isolated, away from the border, and of lower economic significance. Furthermore, since almost all Greeks holding foreign citizenship were registered

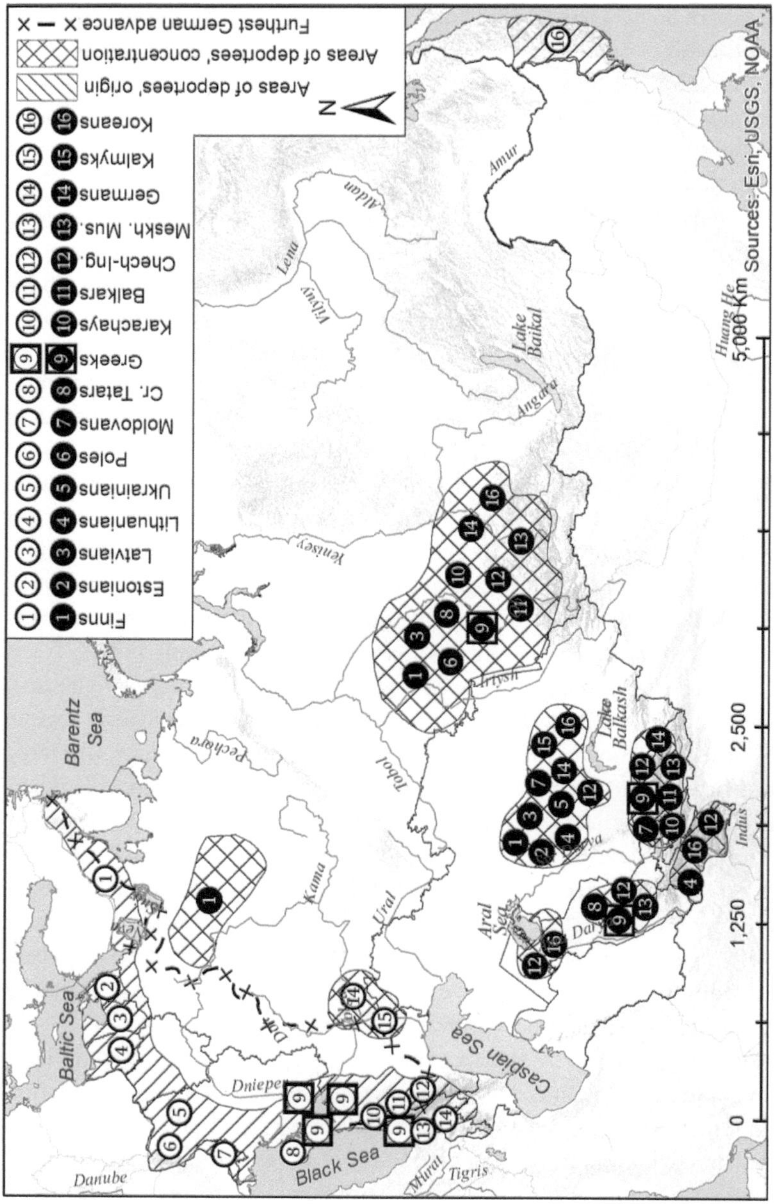

Map 7 Main directions of Soviet mass deportations 1937–1951

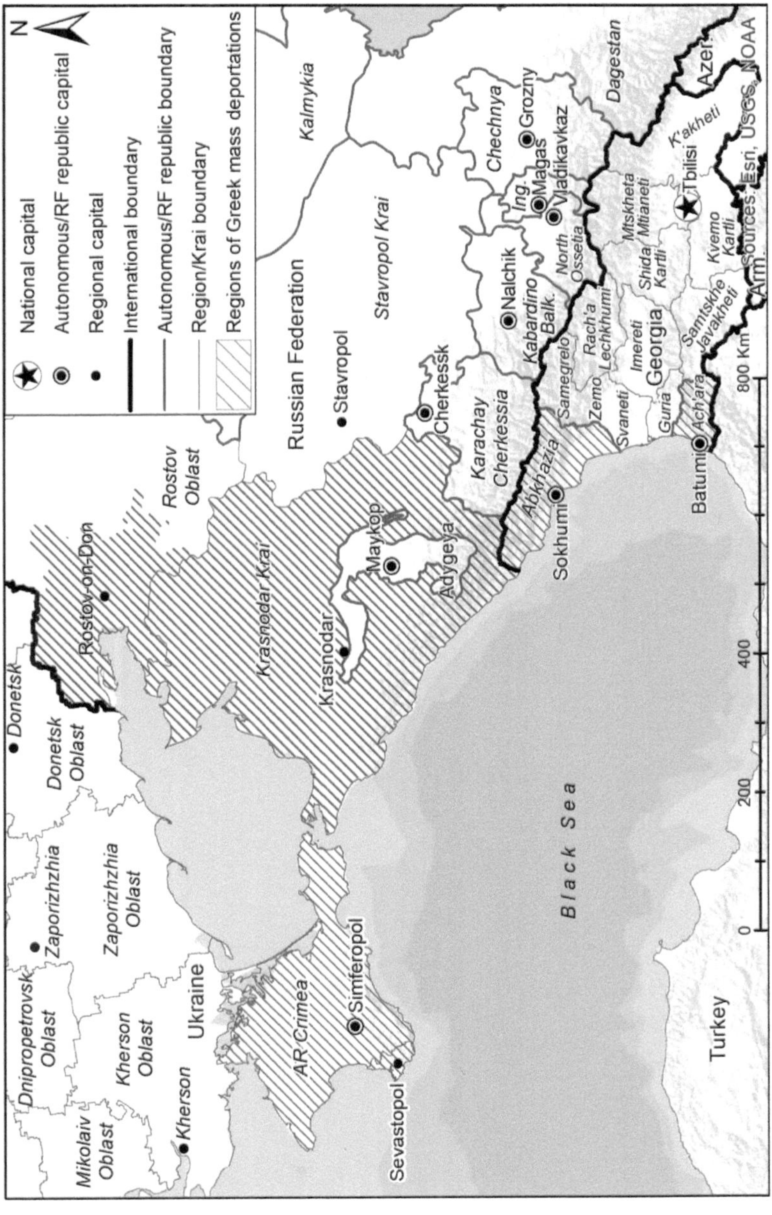

Map 8 Regions of Soviet mass deportation of the Soviet Greeks

in Adjara and Abkhazia, the Greeks of Kvemo Kartli may have been fortunately overlooked in this respect.

Ultimately, Stalinist repression and deportations significantly influenced the demographic structure of Georgian Greek population. As a consequence of forced migration, the ethnic distribution was transformed on both regional and SSR levels. Importantly, these traumatic events strongly shaped the construction of Greek ethnicity in the coming decades. Persecution on ethnic grounds, whereby the regime declared them "foreigners" and brought forward the concept of a "historic homeland", eventually played an important role in post-Soviet migrations (Popov 2016), to which we now turn.

Chapter V. Emigration from Georgia

> *"We note that many who have attempted to escape the Charybdis of methodological nationalism are drifting towards the Scylla of methodological* fluidism."
>
> Wimmer and Schiller (2003; 600).

On the 26th of December 1991, when Mikheil Gorbachev announced the dissolution of the Soviet Union, a civil war was already raging in Georgia. The country entered independence amid war and chaos. For the next several years, Georgia's economic development was stifled by the dissolution of state structures, widespread activity of criminal gangs, and a general environment of corruption and lawlessness. Between 1990 and 1994, gross domestic product based on purchasing power *(GDP (PPP))* decreased by 72.6%, from USD 11,135 to USD 3,045. It required 22 years to get back to the USD 11,000 mark (Figure 9). The decline was even worse than that experienced by the German economy after the Second World War, where GDP (PPP) dwindled by 64%, from USD 6,084 to USD 2,217 (Maddison 2003; 61–62)[26]

The economic collapse that began in the 1990s was accompanied by hyperinflation, and further amplified by the loss of access to former Soviet markets and the abrupt separation from Ruble monetary zone. In 1990, average monthly inflation was 1.7%. In 1992, it reached 21.3%, and by 1994 it had climbed to 56.3%. All in all, hyperinflation in Georgia amounted to a staggering 3310.6% between 1990-1995 – three times as much as in other post-Soviet countries, according to official statistics (Salukvadze and Meladze 2014).

Another important issue was poverty, which affected up to 80% of the population at its peak in 1994. In the following years, although poverty rates declined, they remained severe: 60% in 1995 and went to the lowest 46% in 1998. However, the state budget kept being chronically underfunded to solve this problem and in 1999–2003 the level of poverty has stabilized at around 51%–52%. (GEOSTAT1999:75–77; IMF 2003:11).

At the same time as Georgia went through economic and institutional collapse as well as violent conflict, international borders opened for the first time after 70 years of Soviet rule. In these circumstances, mass labor emigration was only to be expected. Every ethnic group was involved in emigration from Georgia (see table 7). Aaccording to the 2002 population census, ethnic Georgians took up

---

26 In 1990 International Geary-Khamis dollars.

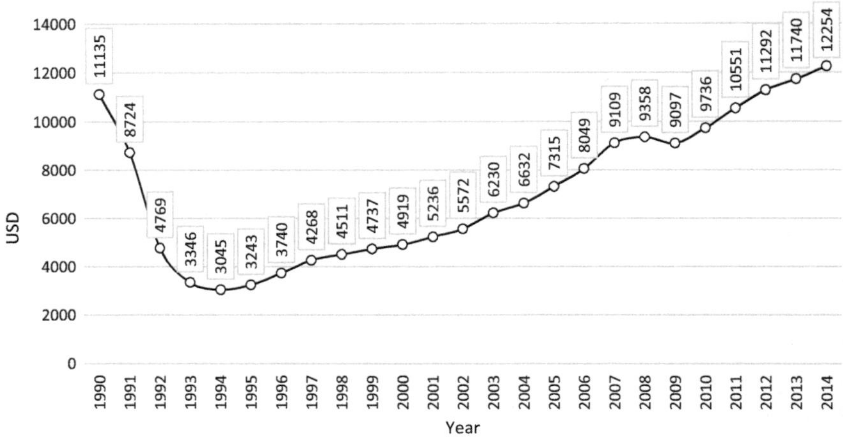

Figure 9 GDP per capita, PPP (constant 2017 international USD) in Georgia, 1990-2014 (Source: World Bank 2021)

Table 7 Net Migration in Georgia, 1990–2014

Year	Number ('000)	Rate (per 1000)	Year	Number ('000)	Rate (per 1000)
1990	-13.2	-2.4	**2003**	-27.5	-6.4
1991	-22.6	-4.1	**2004**	5.5	1.3
1992	-139.3	-25.8	**2005**	76.3	17.5
1993	-140.9	-27.4	**2006**	-12.1	-2.8
1994	-142.6	-29.3	**2007**	-20.7	-4.7
1995	-127.2	-26.9	**2008**	-20.5	-5.3
1996	-123.1	-26.7	**2009**	-34.9	-9.2
1997	-59.9	-13.2	**2010**	-30.4	-8.0
1998	-39.2	-8.7	**2011**	-35.9	-9.6
1999	-36.3	-8.2	**2012**	-21.5	-5.8
2000	-35.2	-8.0	**2013**	-2.6	-0.7
2001	-32.6	-7.4	**2014**	-6.5	-1.8
2002	-27.8	-6.4			

the largest share (60.9%) of emigrants, followed by Armenians (11.1%) and by Azeris and Greeks (7.7% each). However, in the case of the Greek population this led to a dramatic decline. As noted above (section 3.2; Figure 1), compared to the 1989 census the Greek population had diminished by 84.9% in 2002, from 100,324 to 15,100 individuals. In 2014, with only 5,500 Greeks left in Georgia, the overall decline since 1989 reached 94.4% – the largest among all ethnic groups in the country.

5.1 Labor migration: the most important factor

Quantitative and qualitative analysis of the interviews carried out during fieldwork indicates that the mass outflow of the Georgian Greek population is better understood as primarily a labor emigration process than as one of "repatriation" primarily motivated by ethnic or religious forms of belonging.

As illustrated in Figure 10, which is based on the quantitative questionnaire, 56% (22 out of 39) of the valid cases (i.e. those who responded to the question) directly pointed to the economic situation as the main push factor for their family members' migration. Family reunion was named as the most important pull factor and was cited by 11 interviewees as the reason behind the decision to emigrate (four Urum and seven Pontic Greeks). As the respondents explained, the need for family reunion corresponds to the end of the preparation process for emigration. This means that one or several members of the family had already emigrated, and as soon as the financial and legal means were available, the rest of the family left Georgia to join their emigre relative(s).

All interviewees mentioned economic hardship as the reason for sending the first family member(s) to another country. In this regard, 33 (and not 22) of the 39 respondents mentioned the difficult economic situation, high unemployment rates, and low income as the main driving factors for their (family members') emigration from Georgia. No respondents (0) mentioned a desire for "repatriation" in the "ancestral homeland" an important factor in the decision to emigrate.

To summarize Figure 10, the radar chart clearly shows that for both Urum and Pontic Greeks two motives stand out: the economic situation and family reunion.

Importantly, there are no significant differences between the Urum and Pontic populations in the survey. Therefore, in the forthcoming sections, which discus the quantitative data on Georgian Greeks' economic situation as they influenced their emigration process, the two groups are considered without distinction.

The qualitative semi-structured interviews provide a more nuanced representation. When respondents tried to express the reasons for their or family members' emigration, they also mentioned a number of secondary factors. These include ethnic belonging, the nationalist atmosphere after independence in the 1990s, the Greek government's "open door policy", the high demand for cheap

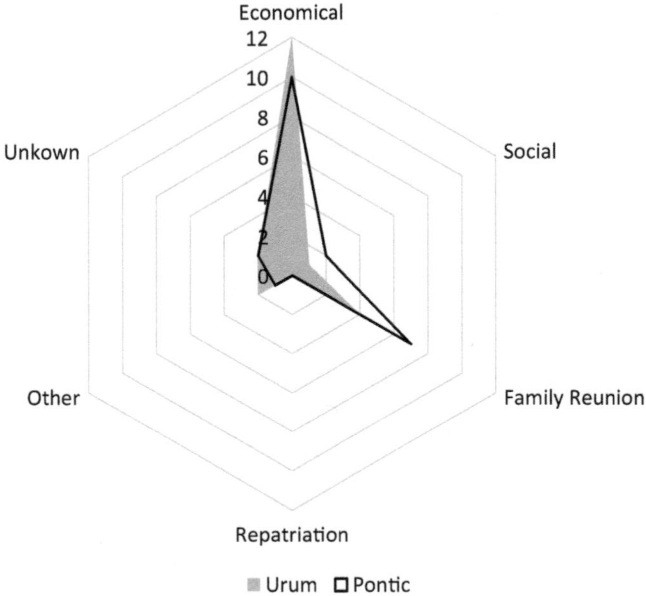

Figure 10 Factors of migration according to interviews

labor in receiving countries, emigration networks, and social-cultural factors (section 5.3).

In most cases, respondents found it difficult to name a single factor. Rather they named several, indicating that they played a concurrent role in the decision-making process around emigration. This supports the relevance of a theoretical synthesis, as suggested in Chapter 2.

For instance, AK from Batumi, who has personal experience in seasonal migration, names the economic crisis and high unemployment after the collapse of the Soviet Union as the main push factors for emigration. AK further noted that later migration was also spurred on by the hopes of family reunion. His parents, who had stayed in Georgia, left to join their children abroad:

> *Greeks probably left because they wanted to live better lives. The last Greeks, the ones who left cities, the people who have left in the last 10 years, their reason for leaving was more to join their children. But apart from that, the majority of Greeks left from villages: those who were working tangerines, tea. After the collapse of the Soviet Union, nobody wanted either tea or tangerines, and these people needed to live and work, didn't they? So, they were leaving for three months to work, then they were coming back during winter, then they were*

leaving and coming back again. They were living like Gypsies, and the result was that they never gained anything, and all the money they were making they spent on feeding themselves. That's why, in the end, they decided to stay there.

AK, 37 years old, Batumi

The story AK tells describes two chronological stages of labor migration: the first one, when one or several separate family members were involved in seasonal migration, and the second, when the whole family was reunited in emigration. This supports New Economics theory in its emphasis on how individual migrants are influenced by the household as a collective actor in economic survival. Later, after seasonal migration proves unsuitable for the migrants, as shown by a simple cost-benefit account, seasonal migration transforms into permanent emigration.

A 65-year-old Urum woman, EM, living in a village in Ts'alk'a municipality expressed the economic narrative clearly. She spoke about one of her family members decision to emigrate as motivated by purely economic reasons. She described the harsh financial situation and the general atmosphere of hopelessness as what led her daughter, who had received higher education in medicine, to emigrate:

They left because Georgia was collapsing. I'm telling you, there were no salaries. Doctors didn't get their salaries in hospitals. I didn't have one either. But we survived, because we had cows. We were planting potatoes. What more is there in Ts'alk'a? So that's why they left, but even if it had been possible to leave during Soviet times, they would never have done so. But our people [Georgian Greeks] found itself in such a situation that they cannot even come back now. They sold their houses for some spare change. I was crying day and night when my [daughter's name] told me – "Mother, what should I do here? There are no jobs." When she finished her studies and became a doctor, what was there for her to do with her salary of 150 Laris? How could a doctor live like that? So, they left.

EM, 65 years old, Village in Ts'alk'a municipality.

While this story focuses on salary and professional options, and frames the decision to emigrate in terms of a cost-benefit calculation, what is noteworthy is that the respondent also emphasises that during the Soviet Union, if there had been the opportunity to emigrate, "we" or "our people" would not have done so. This suggests that, in her opinion, only radical economic problems were important for Georgian Greeks, rather than the group identity signalled by the expression "our people". This strengthens the idea that their emigration was primarily motivated by economic reasons.

5.1.1 (Un)Employment: background and current rates

Unemployment is a fundamental cause of labor emigration and, accordingly, a cornerstone of any migration theory. To this day, unemployment in Georgia remains high, and contributes to the country's negative net migration, as seen in Table 7. For instance, according to official data, in 2018 the negative net migration rate in Georgia was 10,783. The number of emigrants amounted to 98,935 people, 15.8% more than in 2017. In addition, 86.6% of emigrants were of employable age (15–64).

In April 2015, a survey carried out by CRRC-Georgia for the National Democratic Institute (NDI 2015) showed that 67% of the Georgian population considers the lack of jobs to be the most important issue in the country. 66% of the respondents reported being jobless. According to ILO, however, in 2014 the level of unemployment in the labor force was 17.4%.

This considerable difference is due to the incompatibility between unemployment perception and the official methodology(ies) used to assess unemployment. Like almost all other countries, Georgia uses the methodology designed by the International Labour Organization (ILO).[27] Accordingly in the documents of the National Statistics Office of Georgia (2019) an employed individual is defined as:

> *An individual of at least 15 years of age, who has worked (at least for one hour) in the seven days prior to the interview with the objective of receiving an income (salary, in kind, profit and so on), was helping some household member without remuneration, or was not at work for some reason, but was formally registered as a worker.*

Nevertheless, indicators for perceived unemployment should not be ignored when discussing labor migration. In Georgia, citizens found themselves rather suddenly in a free market system after living in the Soviet economy. Therefore, it is often the case that people only consider themselves employed if they have a fixed salary as an employee. Because of this perception, those who consider themselves "unemployed" may include, for instance, taxi drivers,[28] unlicensed

27 The new methodological standard was proposed by the ILO to Geostat in 2013, but officially entered into force only in October 2018

28 Since the day Georgia became independent, driving a taxi does not require a license or to pay any kind of tax. Any individual who owns a driving license can become a taxi driver. The only exception is Tbilisi, where it became necessary to acquire a license on the 1st of October 2018 and to pay a fixed yearly fee. The situation is also mainly unregulated regarding street vendors.

street vendors, people cultivating their own plot of land, and so on. According to official statistics, however, they are considered employed.

This incompatibility means that official statistical data are only one part of the picture. For instance, Figure 11 presents 3 versions of the unemployment rates: 1. according to the Geostat's old methodology, which is incompatible with ILO's international standard adopted since 1982. 2. according to academic researchers' calculated estimates and 3. according to recently published unemployment statistics for 2010–2020, which for the first-time recalculated rates according to the International Labour Organization's methodology (Geostat 2020).

Taking into account the average 9.3-point difference in unemployment rates given by old and new standards, as well as the scale of economic collapse and hyperinflation, the validity of official unemployment rates in the first half of the 1990s is clearly questionable. On the one hand they do not fit the methodological standards, and on the other hand they show no apparent influence of the economic collapse on the labor market. In contrast, according to experts, the level of unemployment rose to 45% of the economically active population in 1992 (Tvalchrelidze et al, 2011:49, Labbate 1999;75).

The difference can also be explained, in part, by the chaotic transition period, during which workers were still formally registered as employees in enterprises and factories which, in reality, were no longer functioning. However, the main reason was the 1992–3 land reform, which distributed parcels of state-owned

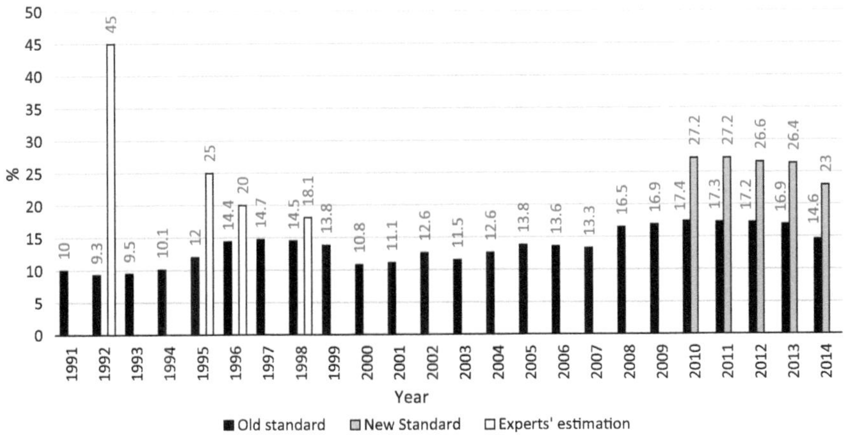

Figure 11 Unemployment in percent of total labor force in Georgia, 1991 – 2014
Source: Geostat; Tvalchrelidze et al, 2011; Labbate 1999.

agricultural land to the population. While the allocation per family (up to 1.25 ha) was too small to offer economic gain, it increased many households' food security during the years of transition. As Salukvadze and Meladze note (2014; 161–162), this artificially influenced labor statistics:

> *Whereas in 1990 the share of agricultural employees comprised 26%, in 2003 this figure grew to 51% ...as hundreds of thousands of families automatically became recorded as self-employed in agriculture. Curiously, the self-employed among made up 70% of the employed population.*
>
> *In reality, the lack of mechanization, the unavailability of fertilizers, seeds and other materials for agricultural production, together with the inaccessibility of credits and the fact that it was predominantly subsistence farming on tiny land parcels, did not allow the majority of farmers to generate sufficient income from their land. Consequently, many of them gave up agricultural activities and "joined" the unemployed population, while others preferred to emigrate from the country.*

Our fieldwork data also show the limits of a categorical distinction between employment and unemployment. Figure 11 displays the number of respondents who consider themselves unemployed but would be considered employed by official statistics. Out of 42 interviewees who gave valid responses, 18 self-reported as unemployed (but wanted an appropriate job and would be able to start one in the next 2 weeks), 17 were employed, six were already pensioners, and one respondent was employed seasonally. Figure 11 also indicates that out of the 18 interviewees who claimed to be unemployed, 12 were involved in farming or cattle breeding – leaving only six who would officially count as "unemployed".

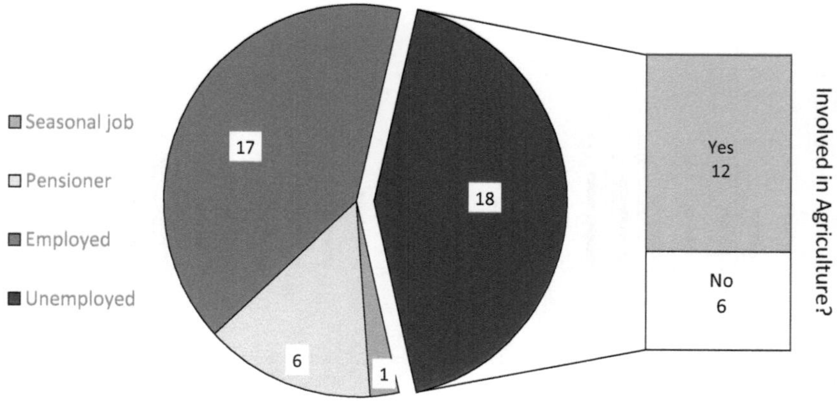

Figure 12 Rate of unemployment among interviewees

More broadly, during the fieldwork it became apparent to us that many people were involved in some form of agriculture, especially in the Kvemo Kartli region. However, they were not selling their produce, and said they used most of it to feed their own family (Loladze 2016). A clear example is the following excerpt from the interview with LP. He is a 28-year-old Urum Greek who is planning to emigrate to Greece, and gathering the documents needed to apply for Greek citizenship. He currently lives in a household of six, in the former home of his emigrated cousin, and also takes care of other emigrated relatives' agricultural land. Nevertheless, he considers himself to be unemployed:

NL: Do you own a plot of land on your territory of residence and, if so, what do you use it for?
LP: Yes, I do. I have three vegetable gardens; 800 square meters here – now this is a real vegetable garden! I have another 800-square-meter vegetable garden here, at my relatives' house, and another one of 800 square meters. That's it. That's all. I do not even have half a hectare, it's not enough!
NL: If you could change your place of residence, would you live in a city or in a village?
LP: To be unemployed in the city will not bring me anything, so I still prefer living in a village, because I'll never be hungry here. As someone said: "Dig, sow, pick, and eat [laughs]!"

LP – male 28 years old, village in Ts'alk'a municipality

In general, the majority of Greeks consider unemployment to be the main difficulty for living in Georgia. During our fieldwork, in response to the question "What are the main advantages of living abroad?", 29 out of 34 respondents named employment and high wages. Correspondingly, to the question "What are the disadvantages of living in Georgia?", respondents named unemployment and insufficient wages. According to the same LP, whose views are shared by the majority of the respondents, emigration would never be considered if employment and reasonable wages were available, as they had been during the Soviet period:

NL: In your opinion, what are the main advantages of living abroad? What are the benefits?
LP: The advantage is that over there there's work. Here, work by itself isn't enough. Otherwise why would people leave this place? We were not living that badly, were we? I'm not saying that wages should amount to 50 or 40 Laris, but if it was 20 per day here [about 9.5 EUR], I would be very satisfied. I would have a very good income. And why would I need to go anywhere else then? Why would I take the trouble of speaking a foreign language? It's not that bad here, is it?

LP – male 28 years old, village in Ts'alk'a municipality

LP's explanation clearly shows how harsh economic conditions act as a push factor, while prospectively better working conditions and higher wages in the receiving country are pull factors. Notably, he underlines the difficulty of speaking a foreign language, being an Urum Greek who received his education in Russian, and only began studying Greek after deciding to emigrate. The value of a theoretical synthesis is shown by the fact that in the following section of the interview, discussed in sub-chapter 5.3, LP underlines that he would prefer, if he had the opportunity, to emigrate to Russia due to fewer cultural/linguistic barriers and his existing circle of friends, in other words, his social capital.

A 33-year-old Pontic Greek woman, GN, is from Adjara. She also explains Greeks' emigration in terms of the harsh economic situation, corruption and unemployment in the 1990s. At the time of the interview, she had already applied for Greek citizenship and was waiting for the result. Apart from GN and her parents, all of her family members, including siblings, grandparents, uncles, and aunts, had received Greek citizenship. Some live in Greece, and the rest elsewhere in the EU. GN herself had been lucky to find a highly-paid position working for a foreign company in Georgia, and did not consider leaving until a few years before the interview, when the company experienced financial problems (see also sub-chapter 5.3). She then decided to apply for Greek citizenship, but says that after the economic crisis in Greece, this became much more difficult:

> *NL: Why did Greeks leave Georgia?*
> *GN: I can tell you right now, it's because it was difficult here. There was no electricity in villages. Because people just couldn't live here anymore. How do you explain to a child that you can't give them bread with salami, so they have to eat just bread? And for that bread you had to stand in line, because it was a coupon system. I remember a close neighbor of ours, who would wake up at six in the morning and stand in line for us. Now, of course, it's not like that, now you can go and buy it whenever you want.*
> *Apart from that, in the first place there were no jobs. Let's begin with that. Not everybody could find a job. There was a clan, the Abashidzes [formerly a local political leader], a few other surnames, it was really like that. So, the situation was that if you had no connections, it was not realistic to hope for a good job. People were in real hardship then. That's why they were leaving, because there were better jobs there, a better life, everything was much better, and it's still like that now. Living standards are much higher in Europe.*
>
> GN female 33 years old, Batumi

As we can see from this interview, GN too considers unemployment to be the driving factor behind emigration. She notes that the imbalance between Georgian and European countries has persisted to this day, with "high living standards" remaining an important pull factor.

5.1.2 Poverty

Closely linked to unemployment, poverty is a particularly important incentive for labor emigration. Since the 1990s, poverty in Georgia has fluctuated as the country's fortunes changed, but assessing poverty rates during the first years of independence is quite difficult. This is because, in the first place, no standard methodology is used across different studies by organizations such as Georgia's National Office of Statistics, the World Bank, the International Monetary Fund and UNICEF (Sulaberidze 1999; Gugushvili, 2011; Geostat2019). Furthermore, the national methodology itself has changed several times. Until 2004 the state used an absolute, subsistence minimum criterion to determine the poverty line. As illustrated in Table 8, this put 80% of the population below the poverty line in 1994, and 60% in 1995. According to this methodology, poverty was at its lowest in 1996–97 (46%), climbing back up to 54.3% in 2003.[29]

In 2004, however, the new government established after the "Rose Revolution" decided to change the methodology, defining a relative poverty line at 60% of median consumption. As a result of this decision, poverty level artificially dropped by half – but as Table 8 shows, over 20% remained in relative poverty between 2004–2014, with a decrease of only 2.3% during that period.

A further change in poverty-assessment methodology was introduced in 2017, when the National Statistics Office of Georgia began publishing an absolute poverty indicator in order to adapt the data to global standards. The absolute poverty methodology is based on the "basic needs approach" introduced by the World Bank and used in more than 100 countries. The basic needs approach determines the poverty line based on the cost of the minimal calorie intake recommended for a man fit for work. According to this methodology, 36.9% of the population had been under the absolute poverty line in 2005. In the following years, this had declined steadily to 21.6% in 2014.

Despite these different methods and the indicators they provide, Table 8 shows that like unemployment, poverty in Georgia is persistent, albeit declining, and therefore remains an important driver of emigration.

Concerning the Greeks, as the interviews showed, low incomes and poverty were major push factors for Greek emigration and remain challenges to this day. Figures 13 and 14 illustrate the reported monthly personal and household incomes of 46 respondents (in EUR, based on the concurrent exchange rate).

---

29 Since 2004, the National Statistics Office of Georgia introduced a correction in the subsistence minimum, and with the new methodology of calculation, reduced the consumer basket (a sample of basic consumer goods) by almost one third.

Table 8 Poverty Levels (%) in Georgia according to different methodologies in 1994–2014

	1994	1995	1996	1997	1998	1999	2000	2001	2002	2003	2004	2005	2006	2007	2008	2009	2010	2011	2012	2013	2014
Sub-sistence minimum	80	60	48.2	46	50	53	51.8	51.1	52.1	54.5	-	-	-	-	-	-	-	-	-	-	-
Relative poverty	-	-	-	-	-	-	-	-	-	-	24.6	24.1	21.8	21.1	22.6	21.6	22.8	23.0	22.4	21.5	21.4
Absolute poverty line	-	-	-	-	-	-	-	-	-	-	34.3	35.4	36.9	38.8	34.9	34.9	37.3	34.1	30.0	26.2	23.5

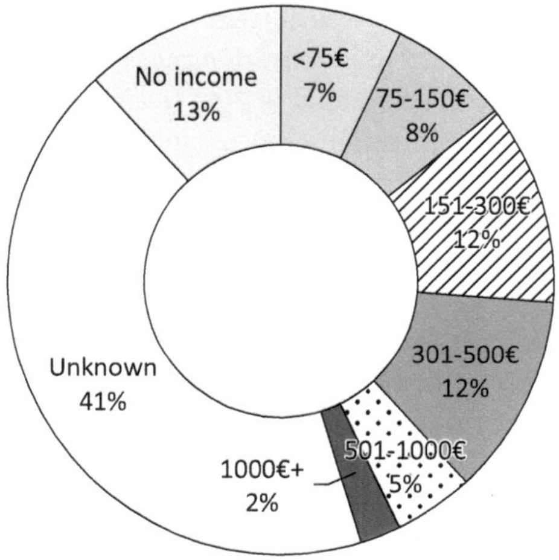

Figure 13 Personal income (monthly) according to interviews

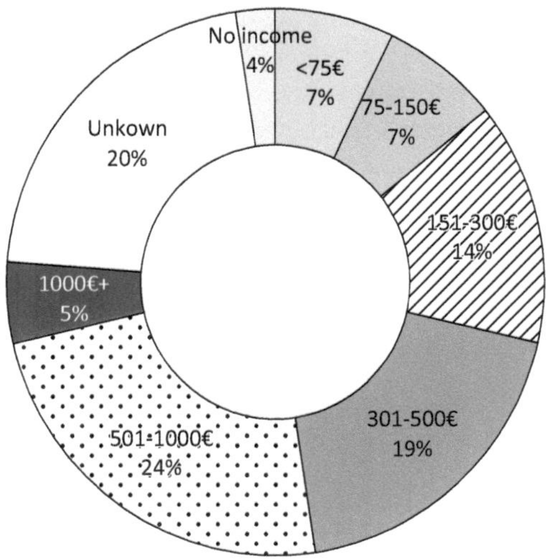

Figure 14 Household income (monthly) according to interviews

As expected, a large portion (41%) of the respondents did not wish to speak about personal income. More than one in nine (13%) said that they had no personal income at all, because they were unemployed at the time of the interview. According to 7%, their personal monthly income was less than 75 Euros. It is important to take this boundary into account, as at the time of the interviews, the subsistence minimum for a male of working age was around 71 Euros (GEOSTAT2014:57). This means that a total of 20% (including unemployed respondents) of the respondents (including the unemployed respondents) said that their income was lower than the subsistence minimum established by the state.

Similarly, the average monthly household income was also quite low (see Figure 14). According to the interviews, only 5% of the families have an income of more than 1,000 EUR. A quarter (24%) reported their monthly household income as below 500 EUR, while 7% said it was between 75 and 175 EUR. Regarding the subsistence minimum, 4% of respondents said their family had no source of income at the time of the interview, and 7% said their monthly family income was lower than 75 EUR.

Figure 15 shows respondents' answers to the question "*What was your average monthly income during the Soviet Union?*". Since it would have been complicated to compare Soviet Rubles with current incomes, respondents were asked choose between five answers: *a) Much more than now; b) More than now; c) The same as now; d) Less than now; and e) Much less than now*. The majority of respondents who were adults and fit for work during the Soviet Union said their income at the time was much higher (21) or higher (2) in comparison with today. Only two respondents said that their income had been lower, and one that it was much lower, during the Soviet period. The same pattern applies to family income.

Figure 15 underlines the economic hardships described in the previous subchapters. Despite Georgia's economic development, poverty and unemployment are still the most significant challenges facing Georgian citizens, Greeks included, and remain the most significant push factors for emigration.

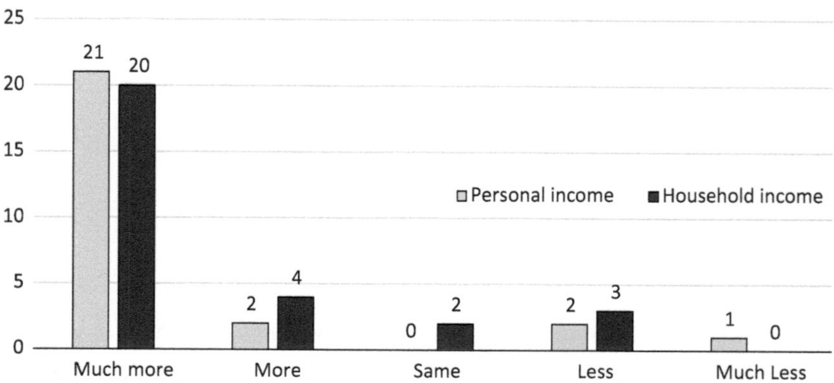

Figure 15 Previous personal and family income in the Soviet Union

5.2 Emigration routes and country comparisons

Even though official quantitative data from 1990–2002 do not allow us to accurately compare Georgian Greeks' emigration vectors with the vectors of overall emigration from the country, several scholars have made such estimates. For instance, according to a 2006 study by the Center for Migration Studies, labor emigration flows from Georgia were mainly directed towards four countries: Russia (52.6% of the total emigration flow), Greece (11.2%), Germany (11%) and the USA (8.2%) (Chelidze 2006; 38–40). These percentages are consistent with 2004 data from a joint study by USAID and Save the Children: 54.5%, 15.6%, 11.1% and 10.9%, respectively (Dershem & Khoperia 2004). From 2002 to 2014, according to official statistics, the five main destination countries were Russia (21.7%), Greece (15.9%), Turkey (11.2%), Italy (10.9%) and Germany (7.1%) (Geostat2016).

Neoclassical, New Economics and dual/segmented labor migration theories all consider the difference between countries' economic environments, directly linked to employment and wage levels, to be the main factor determining emigration. Therefore, this sub-chapter will compare Georgia with Greece and Russia, the two main destination countries for Georgian Greeks, in terms of GDP per capita (PPP), unemployment rates, and average annual wages.

5.2.1 Main vectors of emigration

Analysis of the interviews shows that 32 out of 48 respondents have at least one family member or close relative who has emigrated to Greece, and only a few to other countries – seven to Russia, four to Germany, four to Cyprus, and one to Yemen.

Maps 9, 10, and 11, which are based on the communication networks that exist between the interviewees and Greeks who have emigrated from Georgia, offers more detailed information about the destination countries. The maps illustrate answers from respondents in both Georgia and Greece, when asked to list the countries in which there were Greek emigrants with whom they maintained communication (see Appendix: Semi-structured questionnaire A, Questions 28–28.1). Apart from a mapped visualization of communication networks, this data also provides information about emigration vectors.

Map 9 shows that the Georgian Greeks' main communication links are with Greece and with Russia. As expected, the majority of interviewees (43) named Greece, while 32 named Russia, as the destination country of those Georgian Greek emigrants with whom they maintain communication.

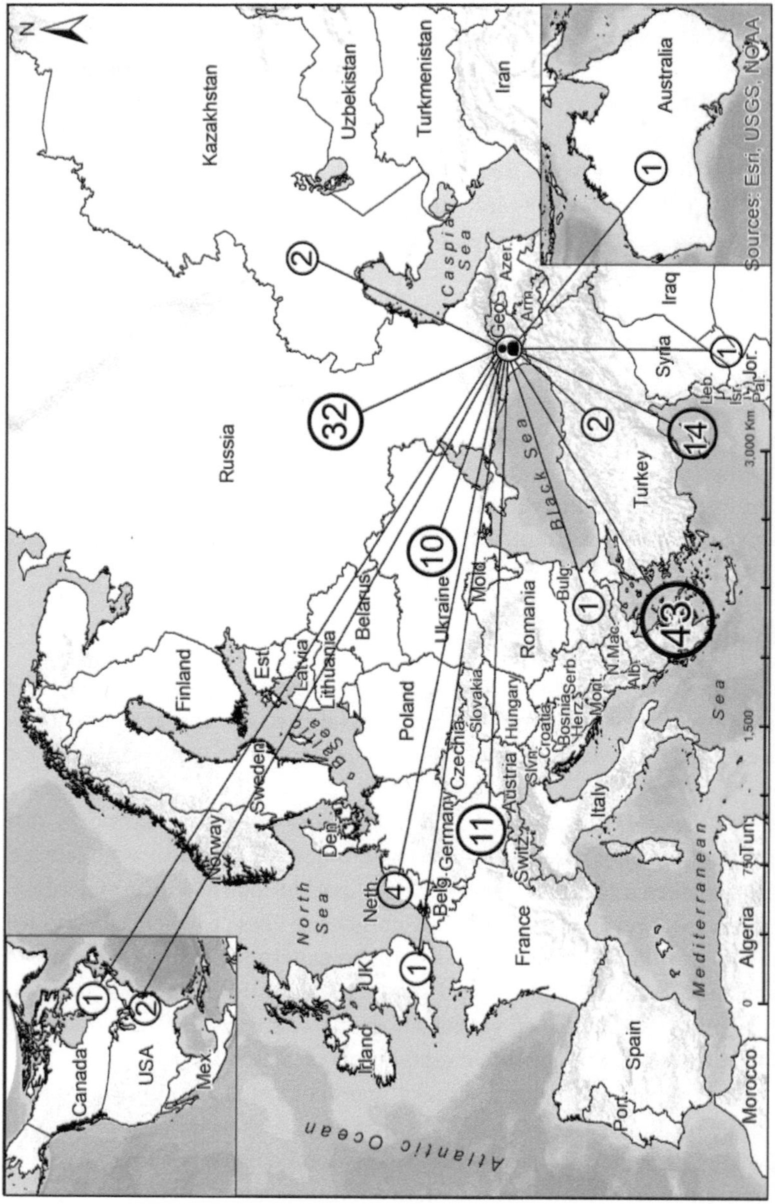

Map 9 Communicative networks/migration vectors of Georgian Greeks according to interviews in Georgia

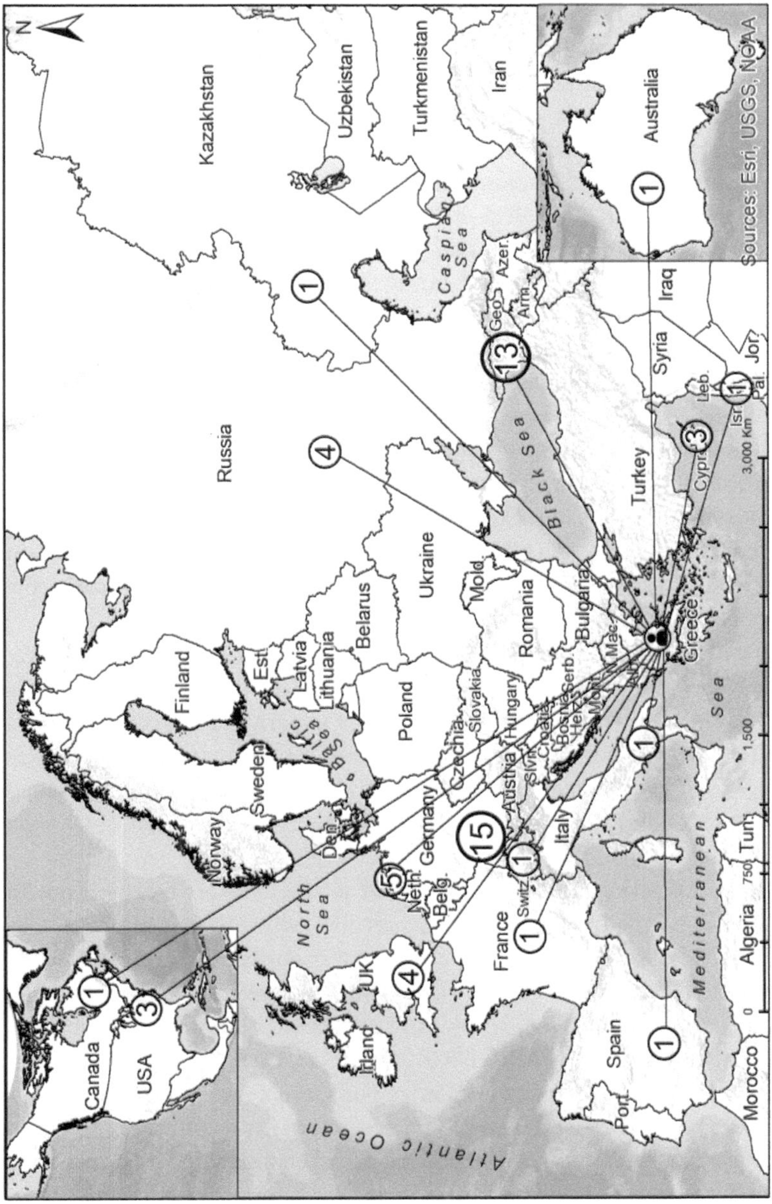

Map 10 Communicative networks/migration vectors of Georgian Greeks according to interviews in Greece

Destination countries also included Ukraine, Germany, and Cyprus. However, only 10–14 interviewees named these. In the case of Ukraine and Cyprus, interviewees might have "forgotten" to include them in the list, because of their perceived linguistic and cultural similarity to Russia and Greece. Indeed, some interviewees' comments indicate that they consider Cyprus to be somehow a continuation of Greece. Ukraine too may be perceived as part of the Russian socio-cultural and economic space.

Other countries were also mentioned, including the Netherlands (5 respondents), Bulgaria (1) and Great Britain (2). A minority of interviewees named the USA, Australia, Kazakhstan and Jordan. With the latter two, they added that emigration to these countries took place during the Soviet period and was related to marriage. Therefore, economic reasons were not their sole cause.

Map 10 offers a visualization of the communicative networks of 25 Georgian Greeks interviewed in Greece with Greeks who have emigrated from Georgia. As mentioned in Section 1.3.3, these interviews were carried out by our collaborator Ekaterine Kartvelishvili during her fieldwork in Greece. Since the sampling was not prepared specifically for this study, and because the small number of interviewees does not allow us to make any sound generalization, the data should be interpreted with caution. Still, it does suggest a number of important migration spaces and socio-economic tendencies (secondary movement from Greece, the privilege of an EU citizenship, etc.).

Compared to the data from our interviews in Georgia, Map 10 shows a geographical spread of destination countries more heavily weighted towards Western Europe. As we can see on the map, the majority of interviewees listed Georgia (15) and Germany (13) in their communicative networks. Georgia was mentioned by 15 of 25 interviewees, apparently a low figure, but one accounted for by the mass migration having left only 5,500 Greeks registered in Georgia in 2014.

The main difference in Map 10 compared to the interviews carried out in Georgia is the geography of destination countries, which shifted more towards Western Europe. As we can see on the map, the majority of interviewees listed Georgia (15) and Germany (13) in their communicative network. Regarding Georgia, 15 of 25 interviewees is low. This can be accounted for by the mass character of the migration, which left 5 500 Greeks registered in Georgia in 2014.

The data related to Germany and other European countries (Cyprus, Italy, France, Spain and Switzerland) shows a tendency towards secondary movement. After emigrating to Greece and acquiring Greek citizenship, migrants were allowed to freely move to other European countries with better economic environments, with Germany holding the lead in this regard. If the links to post-Soviet countries can be explained by a common socio-economic and cultural past, the same cannot be said of countries in the European Union. For instance, when talking about the reasons behind emigration from Greece and Georgia, our interviewee GN discussed just above (sub-chapter 5.1.1) underlines the advantages of European citizenship and social links:

> *NL: Where did they emigrate apart from Greece?*
> *GN: I'll tell you. Many of my family members are in Germany, in Italy. There are also many in Russia. Many also stayed and didn't move from Kazakhstan. In England and Switzerland too. In France too, in Lyon.*
> *NL: Did they move to these countries directly from Georgia or from Greece?*
> *GN: Through Greece, of course.*
> *NL: Could you tell us approximately how long they were living in Greece?*
> *GN: I don't know. I think they left quite soon. I would also have left right away.*
> *NL: Why are you leaving?*
> *GN: I'm planning to leave this summer, and if I can get citizenship there, I'll do it from there. Because if I get it, my child will also automatically become a Greek citizen, and it will serve them in the future. All my relatives and friends are abroad. My child's godparent is in London. Everybody is telling me to join them. You'd agree that it will be easier for me as a European citizen than as a Georgian one.*
>
> GN female 33 years old, Batumi

Map 11 combines the data used in the first two maps and presents a visualization of the main routes of Georgian Greeks' emigration. The data show the two stages of the process, namely emigration from Georgia and secondary movement from Greece.

Emigration from Georgia was mostly directed towards post-Soviet countries (Russia and, to a lesser extent, Ukraine) and towards Greece. In the latter case, the Greek state's migration policy of "repatriation" formed a socio-political pull factor, with the opportunity to easily acquire a European passport playing an important role in triggering this migration vector. In the case of Russia, pull factors were the visa-free regime in force until 2000,[30] personal relationships formed in the past, a relatively strong knowledge of the Russian language among Georgian Greeks, geographic proximity, and familiarity with Russian economic practices and cultural norms.

Greece appears to have acted as a kind of way-station, from which immigrants, after acquiring Greek citizenship, could move on to freely work and live in the European Union. Greek citizens also benefited from visa-free movement to the USA, Canada, and Australia (Hellenic Ministry of Foreign Affairs 2019). Even though Greeks need a visa to enter Russia, secondary movement from

30 Russia unilaterally introduced a visa regime for Georgian citizens in December 2000.

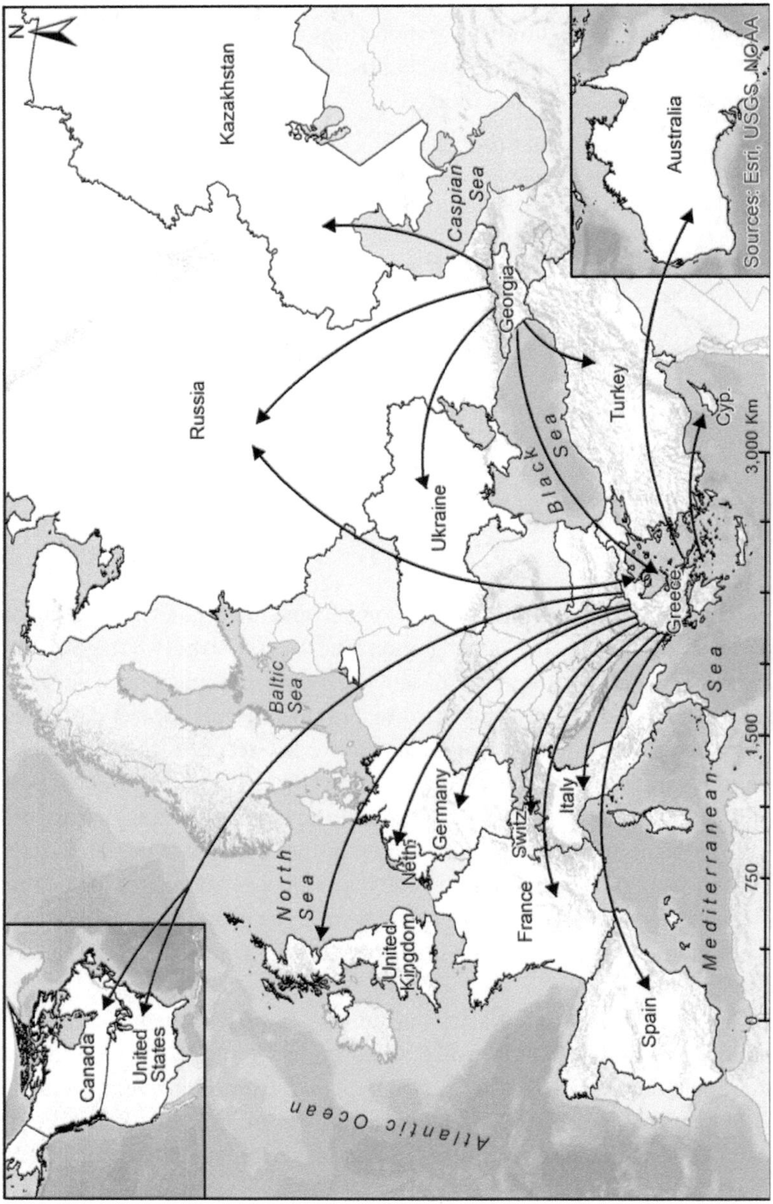

Map 11 Main emigration routes of Greeks from Georgia

Greece to Russia can still be noted, induced by personal relationships and the financial crisis that struck Greece in 2007–2008. The specifics of these processes are discussed in more detail in subchapter 5.3.

5.2.2 Comparison of GDP per capita (PPP)

Figure 16 illustrates the total GDP per capita in Georgia, Greece, and Russia between 1990–2014, in terms of purchasing power parity. It clearly shows that Georgia lags behind. In the period of Georgian Greeks' mass emigration (1991–2002), Greece's GDP per capita was, on average, 480% greater. In 1990, when Georgians were still counted as Soviet citizens, the gap had been 187%, but shot up to 716% in 1994, slowly declining since.

Between 2003–2014, the Georgian economy began to stabilize and recover. During this period Greece's GDP per capita in PPP terms was on average "only" 262% greater, in large part due to the Greek economic crisis of 2007–2008. Nonetheless, the gap remained significant. For instance, in 2014, although the Greek economy was going through another large-scale crisis, its GDP per capita amounted to 28,616 USD, compared to only 12,255 USD in Georgia.

In Russia, despite a difficult transition period following the collapse of the Soviet Union, the economy and living conditions measured by PPP were still much better than in Georgia – although not as much so as in Greece. For example, during 1991–2011, Russia's GDP per capita averaged 218% that of Georgia. The smallest and largest gap were again recorded in 1990 and 1993.

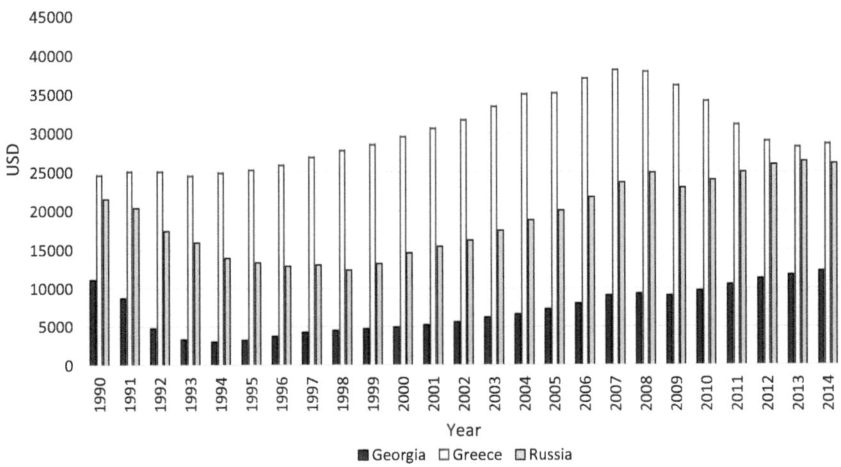

Figure 16 GDP per capita, PPP (constant 2017 international USD)

It is worth noting that unlike Greece, the economic gap between Georgia and Russia remained relatively stable in 2003–2014. During that period, the statistical wage of a Russian citizen was 149% larger than that of a Georgian citizen. According to 2013 data, GDP per capita in Russia was 26,413 USD, the closest it ever came to that of Greece (28,217 USD). This should be considered an important pull factor for Georgian Greeks' secondary movement from crisis-ridden Greece to Russia, together with the aforementioned social-cultural and linguistic factors.

5.2.3 Comparison of labor markets

The comparison of unemployment rates for Greece and Russia illustrated in Figure 17 is based on International Labour Organization data (excerpted from World Bank 2020b). For Georgia the rate is based on the sources used for Table 11, and combines the rates given by the new (2010–2014) and old (1991–2009) methodological standards, as well as experts' estimations (1990, 1995–1997). This clearly shows that up to 2013, unemployment rates in Georgia remained higher than in Greece, despite the artificial reduction of the former discussed in sub-chapter 5.1.1. Unemployment in Greece significantly increased only after the devastating 2008 financial crisis, acting as a push factor for secondary movement.

In Russia, according to official statistics, unemployment rates were always lower than in Georgia, except for a brief convergence in 2000. Despite having a lower GDP per capita than Greece in PPP terms, between 1991–2009 Russia also

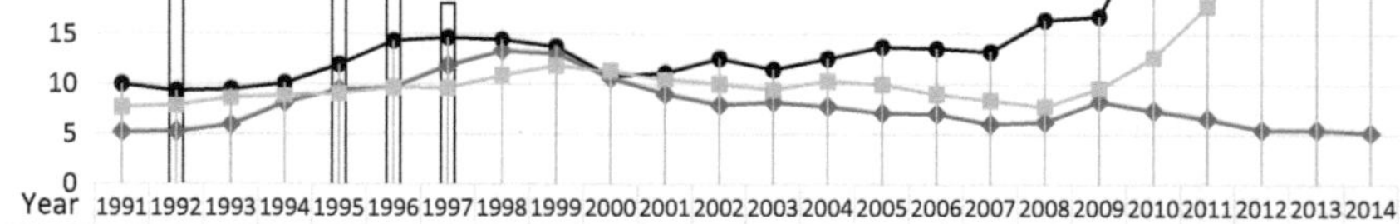

Year	1991	1992	1993	1994	1995	1996	1997	1998	1999	2000	2001	2002	2003	2004	2005	2006	2007	2008	2009	2010	2011	2012	2013	2014
GEO Est.		45			25	30	18																	
GEO	10	9.3	9.5	10	12	14	15	15	14	11	11	13	12	13	14	14	13	17	17	27	27	27	26	23
RUS	5.1	5.2	5.9	8.1	9.4	9.7	12	13	13	11	9	7.9	8.2	7.8	7.1	7.1	6	6.2	8.3	7.4	6.5	5.4	5.5	5.2
GRC	7.7	7.8	8.6	8.9	9.1	9.7	9.6	11	12	11	10	10	9.4	10	10	9	8.4	7.8	9.6	13	18	24	27	26

Figure 17 Unemployment as a percent of total labor force in Georgia, Greece and Russia

had lower unemployment figures, with the exception of 1998, following Russia's financial crisis.

All in all, Figure 17 shows that over the two decades between 1991–2011, even with less than reliable official data, Russia or Greece had significantly higher employment rates than Georgia. This was an important pull factor for labor emigration, a fact equally significant for neoclassical, New Economics, and dual/segmented labor market theories.

5.2.4 Average wages

For the period between 1990–1995, it is quite impossible to construct a statistically valid comparison of average annual wages in Georgia, Russia and Greece, due to the economic turbulence and chaotic currency policy in the former two countries. The Soviet Ruble remained in use in Russia until 1991, before being replaced by the Russian Ruble in 1992. In Georgia, a transient national currency – the Georgian Coupon – was introduced into circulation alongside the Ruble between 1993–1995. The Georgian Lari was introduced on 2nd of October 1995, providing a stabilizing effect (Wang 1999).

From 1995 on, as illustrated in Figure 18, average annual wages in Georgia and Greece differed greatly. Up to 2002, wages in Greece were on average 40 times higher than in Georgia. During this period, the largest gap was in 1995, when the average annual wage in Georgia was equivalent to 129.6 USD, compared to

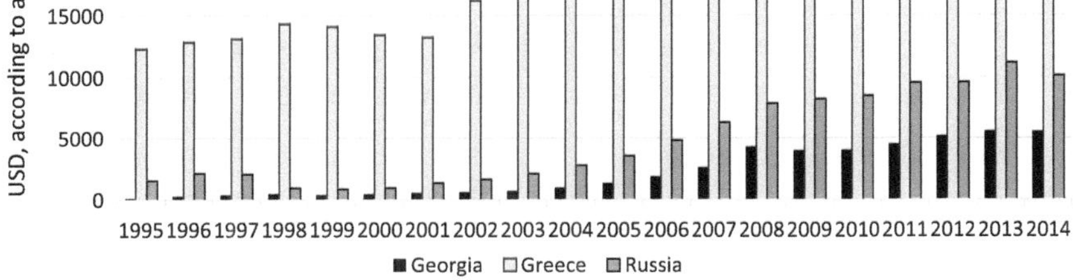

Figure 18 Average annual wage in Georgia, Greece and Russia, 1995 - 2014
(Source: Geostat, OECD 2019, Federal State Statistics Service of Russia 2019)

12 321 USD in Greece. The smallest gap was in 2001 (548.4 USD compared to 13,231 USD).

From 2003, while Georgia's situation improved significantly, the gap with Greece remained large. According to data from 2003–2014, the average annual wage in Greece was 11.3 times higher than in Georgia. The largest gap was recorded in 2003, a difference of 20,026 USD, and the smallest, in 2014, was still large (16,728 USD).

Comparing Russia and Georgia, the difference in terms of average annual wages is both significantly smaller and more stable. For instance, during 1995–2002, the average wage in Russia was about 4.6 times higher than in Georgia. The largest gap was in 1995, with Russia's 1,434 USD being 12 times the Georgian average wage. The smallest gap was recorded in 1998, against the backdrop of a financial crisis in Russia, when its average annual wage was 962.6 USD compared to 478 USD in Georgia.

Between 2003–2014, the difference between average annual wages in Georgia and Russia declined. In 1995–2002, wages in Russia were 4.6 times higher than in Georgia. By comparison, wages were on average 2.2 times lower than in Russia between 2003–2014.

In summary, the difference between Georgia and the main destination countries in terms of GDP per capita, unemployment, and the average annual wage undoubtedly contributed to mass labor emigration of Greeks from Georgia. The fact stands out that even in 2002, after the main phase of the post-Soviet crises, the average annual wage in Georgia was less than 440 USD, compared to 16,167 USD in Greece. Therefore, it is clear that despite the expenses and cost of living associated with emigration, the differences were large enough to be strong pull factors. That said, apart from economic factors there were important cultural, social, linguistic, and political factors which influenced the emigration process and the choice of Greece and Russia as main destination countries. These are discussed in the following sub-chapter.

5.3 Ethnicity and socio-economic ties: aspects of privilege and (dis)advantages in origin and destination countries

Although none of the interviewees identified ethnicity and its concomitant advantages and disadvantages as a driving force behind emigration, the interviews did clearly demonstrate its significance to their experiences.

When discussing their ethnic affiliation, many Georgian Greek interviewees asserted that while living in the Soviet Union they had never experienced prejudice on ethnic grounds. Even today, many said, they do not feel subject to discrimination. However, when addressing recent history, most noted the increase in nationalism during the first years of Georgia's independence and Gamsakhurdia's presidency. Additionally, inhabitants of Kvemo Kartli spoke about the conflicts that emerged in the region following the settlement of Georgian environmental migrants in the late 1980s. Therefore, respondents did consider their ethnic affiliation to have become a disadvantage in the 1990s, compared to the Soviet years.

Concurrently, ethnic affiliation became an advantage in the context of Greek state policy, as noted above in sub-chapter 5.2.1. Against the background of severe economic conditions, this put Georgian Greeks in a position of privilege (Voutira 2004).

Nonetheless, it should be noted that the first generation of Georgian Greeks who emigrated to Greece found it quite hard to assimilate due to linguistic and cultural differences (this was especially true for Urum emigrants). According to the interviews, sometimes the local Greek population did not recognize them as "true Greeks" (Höfler 2020). However, this qualification of their ethnic advantage did not affect their citizenship and could not significantly reduce ethnicity as a pull factor to Greece.

As for emigration to Russia, while there was no explicit privilege for Georgian Greeks, they did have the advantage of competency in the Russian language, as a result of Soviet school-language policy over the previous decades (Höfler 2020: 35–6).

5.3.1 Post-Soviet nationalist challenges in Georgia

In the early 1990s, as nationalist politics took hold in Georgia, discrimination became a factor driving the emigration of ethnic minorities. In some instances, the Gamsakhurdia government's nationalist rhetoric led to acts of harassment. The slogan "Georgia for Georgians" became popular, and non-Georgian populations were often referred to as "guests". The harassment did not reach mass scale, although, nationalist organizations forced Azeri and Ossetian people, particularly in Kvemo Kartli, to abandon or sell their homes and to emigrate (Komakhia 2008; Gachechiladze 2008). This is not reported for the region's Greek population, however.

> 64-year-old AM, whom we interviewed in Tbilisi, spoke about these times:
> *CH: And was it different for ethnic Georgians, Armenians [other nationalities] during the Soviet Union?*
> *AM: No, there weren't, say, [difficult] interethnic relations in the time of the Soviet Union. Nobody cared who got married to who, and people lived together like relatives or whatever.*

There weren't any borders or fences between the two families. But now, now came this "Georgia for Georgians", "Armenia for Armenians", "Azerbaijan for Azerbaijanis". Differences arrived. They tell you what to do. For example, they said to me personally: don't speak Russian! I was talking to a friend coming out of a shop, and I said, apart from Russian I know five languages – so tell me what you can offer me except your Georgian language, but you mustn't behave like that, you mustn't, you understand?
NL: When did this happen? Nowadays or…
AM: No, not now, not now. That was then, during the Gamsakhurdia period. Back then it started. In those Gamsakhurdia times, they started to raise the national question of "Georgia for Georgians". Well, now [recently] I can't say I've felt this personally, myself.

AM female, 64 years old, Tbilisi

AM explicitly notes that during the Soviet period, there was no tension between ethnic groups. On the contrary, she tries to paint a picture of amicable relations. In her opinion, ethnic and national belonging developed meaning only "now", referring to the immediate post-Soviet period. As for discrimination, she tries to isolate it from a different "now" (recent times) and remarks that it only occurred during Gamsakhurdia's presidency.

GA, a 32-year-old male Urum emigrant, who had temporarily returned from Cyprus to his native village in the Ts'alk'a region, spoke to us about the rise of nationalism:

GA: We lived fine until the 1990s. After the 90s and the collapse of the Soviet Union, everybody went somewhere. Mostly they went to Russia, then to Greece. For instance, I've been living in Greece[31] *for 15 years already. I came back a little while ago, about a month ago, but I live there.*
CH: Why did Greeks leave Georgia?
GA: Why did they leave? Because in the 1990s, when there was a war, then he, what's his name, Gamsakhurdia, said that "Georgia was only for Georgians". There was Nazism, that's why. There was no work anymore. There was nothing, and everybody went wherever they could. What do people basically want? Money and a good life – and life wasn't good here. Greece helped us, and gave us documents. I'm a Greek citizen now. Our people [Georgian Greeks] live well there. For example, unlike there, here there's nothing to do. That is why I left, myself. I couldn't do anything here. Others left for the same reason. At first, all our people worked as construction workers. Now they each have their own job, some in the service industry, some have a taxi, some have their own company. Everybody's busy with something, they don't do menial work anymore.

31 Despite mentioning several times that he lives in Cyprus, GA (like other interviewees) often identified Cyprus as "Greece", referring to the wider linguistic and cultural space to which it is perceived to belong.

> *You know, nowadays our people have become real Cypriots there: everybody has a house, a car, a bank account and so on. So, they're like local Cypriots. For example, in the beginning I was working on a construction site. Then I learned a little bit of the language, then this, then that. I got to know some people, as well... But then when Romanians and Poles and others started coming over, everything went wrong. Because while I worked for 120 dollars a day, they started working for 20 dollars. So of course, the owner preferred to employ 10 of them for 20 dollars a day, rather than just employing me.*
>
> GA, Male, 32-years-old, Ts'alk'a.

In his account of motives for emigration, GA mentions the nationalism of the early 1990s, going as far as to use the term "Nazism". However, he soon moves on to economic factors such as unemployment and income. He also emphasizes the Greek government's assistance, indicating the ethnic privilege mentioned above. GA then discusses features that strongly recall the insights of dual/segmented labour market theory: the "menial work" performed by the first generation of migrants, which he says became uncommon over time as they learned the language and found "their own" jobs, and the experience of competition with lower-waged labor migrants from other countries.

Apart from GA, none of the interviewees cited ethnicity-based conflict as a main cause for emigration. While there were and still are conflicts, especially in the Ts'alk'a region between Greeks and environmental migrants, many of our respondents were keen to play down their ethnic dimensions. Instead, they characterized the friction as arising from "everyday life", involving disputes over matters of agricultural land borders and/or property left behind by emigrants.

Nevertheless, it is clear that the wave of nationalism in the first years of independence occupies a significant place in Georgian Greeks' memory, especially the slogan "Georgia for Georgians" attributed to Gamsakhurdia. Hence, ethnic positioning can still be considered a secondary driver in the choice to to emigrate.

5.3.2 Greek migration policy and "privileged labor migration"

Until the 1980s, Greece had a consistently negative migration balance, with the majority of emigrants heading to Germany, Belgium, the USA and Australia. However, from the beginning of the 1990s, geopolitical changes and a significant increase in the demand for cheap labor made Greece a destination for immigrants. Initially, these came from the post-Soviet republics and central Europe, and later from Southeast Asia, the Middle East and Sub-Saharan Africa (Iashvili 2018). Despite lagging behind other southern EU countries on multiple indicators in the 2000s, the Greek economy had the largest share of immigrants in its workforce – up to 10%, according to several estimates (King et al. 2000; Voutira 2004).

In this period, the Greek government decided to support the immigration of ethnic Greeks from post-Soviet countries, a policy motivated not only by the demand for cheap labor but also by ethnic and electoral considerations (Voutira 2004; Kaurinkoski 2010).

The ethnic component involved a desire to reduce the labor immigration rates of non-Greek ("*allogenic*") populations, mainly Albanians, who were deemed a threat to the nation's social and cultural unity. To counterbalance this, the co-ethnic ("*homogenic*") Greek Orthodox populations of the former Soviet Union were considered an important reserve (Pratsinakis 2014). Critical in this context was the "repatriation" and rural resettlement program, implemented in 1991–1999 and co-financed by the Greek state and the EU. According to this program, Greeks "returning" from post-Soviet countries were to be compactly resettled in the northern region of Thrace, which borders Turkey and whose population is chiefly composed of Muslim Turks. This program was ultimately ineffective, as post-Soviet urban immigrants could not adapt to a rural environment, where despite the state's considerable support there were fewer economic opportunities than in Greek or Cypriot cities (Kaurinkoski 2010; Sideri 2006).

In terms of electoral politics, both the liberal-conservative New Democracy party and the social-democratic PASOK, which defeated it in the 1993 polls, sought electoral dividends from supporting post-Soviet Greek immigration. To mobilize votes during the national elections as well as local contests, both parties helped fly "Soviet Greeks", who had the right to vote, from Cyprus and the post-Soviet republics to their (new) voting constituencies (Kaurinkoski 2008;2010).

Under Greek government policy, "repatriated" post-Soviet Greeks received citizenship, residence permits, or tourist visas through a simple bureaucratic procedure. Those seeking to "return" to Greece were only required to apply to the Greek consulate, submitting relevant documents which confirmed that they were Greek, and would receive the desired visa. Such documentation was easy to obtain, since under the Soviet Union all personal documents (ID/Passport) noted the individual's "nationality" (Slezkine 1994; Arel 2003, 2006; Höfler 2017). Additionally, the bureaucratic apparatus in Greece was not equipped to identify forged documents, which were used by many non-Greek emigrant hopefuls in the early post-Soviet period (Pratsinakis 2014; Voutira 2003).

As mentioned above, a state support program was also on offer for Greek immigrants. The program initially helped them resettle in Thrace. However, in 2000–2005, after it became clear that the compact resettlement program was unsuccessful, the Greek government decided to issue low-interest mortgage loans of up to 60,000 Euros, helping up to 35,000 "returning migrants" to purchase land, houses, or apartments in Greece (Kaurinkoski 2010).[32]

Apart from low-interest loans, "returning migrants" enjoyed other integration benefits such as Greek language and vocational training courses; support for agricultural, small and medium-sized business enterprises; and discounts on medical services even to those without Greek citizenship. There were also special quotas for university admission and for jobs in the public service. These benefits were particularly substantial in Eastern Macedonia and Thrace, and despite the ultimate failure of resettlement they still constituted an important pull factor for immigration, privileging Georgian and other post-Soviet Greeks over other labor migrants (Voutira 2003; Diamanti-Karanou 2003; Kaurinkoski 2008, 2010).

The fast track to citizenship for "returning migrants" lasted until May 15, 2001, when the Greek parliament approved a new law making the process much stricter and prolonged. The new provisions concerning "Entry and stay of aliens in the Greek state" and "Attainment of Greek citizenship through naturalization and other provisions" directly impacted post-Soviet Greeks. The law now required applicants to collect archived documents in order to prove their ancestors' Greek identity. Then, depending on the decision of a committee which could take several years, the applicant would be invited to an interview, where their knowledge of the Greek language and Greek traditions, history, and current events would be assessed. After this stage, over a further period of three to six years, another committee would continue to examine the application until a decision was reached (Voutira 2003). Labor emigration via citizenship thus became much more challenging, and could take up to seven years, significantly reducing its attractiveness for potential emigrants in dire economic conditions.

These changing procedures were well-known to our interviewees, either from personal or family members' experience. For example, LP was planning to obtain Greek citizenship, and was collecting the documents at the time of our interview:

NL: What is needed for obtaining Greek citizenship? Is there any type of program that helps you do it?
LP: First of all, there is no program. Secondly, it is almost impossible to obtain [Greek] citizenship. For example, if I were to become a citizen, I would have to go to the local

32 The price of credit differed by region: "In zone 1, covering Eastern Macedonia and Thrace and the North Aegean Sea islands, migrants received 30% of the loan for free and paid no interest, in zones 2, 3 and 4, out of which the latter included Athens and other large cities, the state took upon itself 80%, 60% and 40% of the interest rate, respectively" (Kaurinkoski 2010:126).

consulate, submit an application, wait three to four years before some kind of committee is assembled and begins to discuss my application. Afterwards, I will also be summoned, and after that, they will send my documents to Greece, where, somehow, after several years, someone will make a decision as to whether to give me citizenship or not [laughs].
NL: And how was it earlier?
LP: Previously, any of the local Greeks, and many non-Greeks as well, found specific documents, something, just translated it – where it was written that their parents or grandmother/grandfather were Greek [laughs]. They [the consulate] would discuss it and they would get the citizenship. They mostly gave everyone a positive answer. Negative responses were very rare. Nowadays, for example, my aunt submitted documents and got a negative response for "not having a Greek mentality". Yes, that's how they replied. Now they have such an explanation, but it wasn't like that in the 1990s. Now, the committee requires knowledge of the Greek language, culture, history, and so on, and looks for a reason [to reject someone].

LP – male 28 years old, village in Ts'alk'a municipality

As we can see, LP tried to thoroughly explain the procedure currently required for obtaining citizenship and its concomitant difficulties. He then compared them to the 1990s. In the interview, his sense of defeat becomes apparent, and he does not express high expectations for acquiring legal citizenship. This pessimism is obvious when he speaks about the time required for the procedure and laughs about having to wait several years for the committee's decision. Furthermore, he considers the new requirements regarding "mentality" to be arbitrary and part of a hostile scrutiny of applicants. As a whole, a sense of unfairness clearly emerges from his account, given the relative ease with which the first flow of emigrants obtained citizenship. Compared to LP, they were in a privileged position – as he states, some of them were not even Greek.

Finally, it should be noted that despite the privileges discussed here, many Georgian Greeks still found it difficult to establish themselves in a segmented labor market. For example, AL, a 59-year old woman from a villages in the Tetrits'q'aro municipality, spoke to us about her unsuccessful emigration experience. AL received higher education in Russia, and before the collapse of the Soviet Union she worked in her profession. However, in the early 1990s, unemployment and economic hardship drove her to attempt to settle in Greece:

NL: Where did you work in Greece?
AL: No, I didn't work in Greece. I was only there for two months. They offered me some work, but you know quite well what kind of work. I had to work as a servant, and then go back. Also, I wasn't feeling well, probably because of the climate, so I got up and came back.
NL: Did you return only because of the climate, or were there also other reasons?
AL: The reason was that I hadn't intended to work in that kind of job. I was hoping to work in my own profession. Why would I want to work as a cleaner somewhere? I didn't want

> *that. The fact that I had to work for somebody, clean, take care of them, and so on, despite my higher education, was killing me morally. I couldn't do it. Some people calmly agreed to it and work to this day – they take care of grandmas, they clean. I couldn't do it. Then it so happened that there was a vacancy for an accountant here. The salary is very low, but I can say that I'm satisfied morally.*
>
> AL, female 59 years old, village in Tetrits'q'aro municipality

For AL, it was impossible to adapt to the segmented labor market, with demand for menial workers far exceeding that for professionals. As a result, despite her privileged precondition, AL opted to return to Georgia and work in her profession, lower wages notwithstanding.

5.3.3 The Russian vector and the importance of socio-cultural background

> "Britain had an empire, but Russia was an empire."
>
> Geoffrey Hosking (1997: 41)

As noted at the outset of this book, after Georgia gained its independence, the Russian Federation became the major immigration destination for Georgian citizens and particularly for minorities. This vector is fruitfully assessed in the framework of world systems theory, noting both linguistic and psychological factors (Komakhia 2008).

The Soviet Union, although a multiethnic and multicultural state, was dominated by the Russian language – as a *lingua franca*, a school language, and the major language of cultural and scientific production. This made competency in Russian crucial to upward social mobility, and its status had a significant ideological impact on the majority of Soviet citizens, forming their sense of identification with Russian language and culture (Grenoble 2003).

Our fieldwork data further indicates the primacy of the Russian language for Georgia's Greek population. Among 46 respondents, 44 obtained school education in Russian, one started studying in Russian but graduated in Georgian, and only one interviewee was educated entirely in Georgian.

Unlike the relatively narrow economic focus of the labor migration theories discussed so far, world systems theory incorporates broader historical, cultural, and political considerations. These are important, given the specifics of both the imperial and Soviet state systems. World systems theory emphasizes the colonial and postcolonial dimensions of capitalism, and the interconnection of the metropole and the periphery. Some researchers, working within the traditional Western conception of *direct colonialism*, have excluded the Russian Empire and Soviet Union's distinctive form of territorial expansion from this framework, specifically denying that the USSR should be considered an empire (Suny & Martin 2001). However, as Edward Said reminds us,

> *In our time, direct colonialism has largely ended; imperialism, as we shall see, lingers where it has always been, in a kind of general cultural sphere as well as in specific political, ideological, economic, and social practices. Neither imperialism nor colonialism is a simple act of accumulation and acquisition. Both are supported and perhaps even impelled by impressive ideological formations that include notions that certain territories and people require and beseech domination, as well as forms of knowledge affiliated with domination[...]*
>
> Edward Said (1993: 9)

In this regard, it is entirely possible to speak of a Soviet Empire, whose governance principles reconstructed the classic Russian imperial scheme of the 19th century, but substituted the direct imperial approach with the Bolshevik ideology and a positively-charged "language of friendship and cooperation of the peoples of the USSR."[33] Thus, with the exception of the brief period of independence in 1918–21, Georgia existed under a specific brand of imperialism and colonialism from 1800 to 1991.

The key difference between Russian-Soviet and Western imperialism, according to Susan Layton (2005), is that the former could not unambiguously designate the colonized as an "other", since it was a continental rather than overseas empire, with Asia (to which the Caucasus also belongs) forming an integral part of its territory and history. In this vein, it could be argued that despite being part of a colonial system, it was only in the 1990s that Georgia experienced the extreme deprivation and the position of the kneeling man which Franz Fanon (1961) associates with colonialism. In this respect, Georgia found itself in such a situation only after exiting the Soviet Union, with its lack an independent economy or state institutions itself being a legacy of the colonial system.

The deep cultural-economic factors, and the periphery-metropole connection which formed during this period, influence labor migration to this day. The lack of language barriers, a familiar socio-cultural environment, geographical proximity, and traditional economic ties are essential supporting factors for emigrants.

33 "Language of Friendship and Cooperation of the Peoples of the USSR" was a slogan and the title of the all-Union conference in Tashkent (22–24 May 1979), which announced the new objective of "attainment of complete bilingualism in the Soviet Union" (Solchanyk 1984:23).

LP, for example, despite trying to obtain Greek citizenship as mentioned above, also considers Russia a desirable destination for emigration and permanent residence:

> *NL: In which country would you like to live if you were given the opportunity?*
> *LP: Russia.*
> *NL: And why Russia exactly?*
> *LP: Because there are many jobs there, and Greeks live there as well. A lot of our Greeks went there too. Even from Greece, those who had money in their pockets, all ran over there [to Russia]. They didn't run, but they went there to start some kind of business and then slowly rise to the top. Those who didn't have money came back [to Ts'alk'a]. Some came back to our village, some to other villages as well. As they say, it was a forced return. Some had land here, others a house... You know, there's a saying: "Wait for me here, ugly. I'll find someone pretty, but if I can't, I'll come back to you" (laughs).*
>
> LP – male 28 years old, village in Ts'alk'a municipality

While discussing his decision to emigrate to Russia, LP, similarly to other respondents, emphasized the importance of social contacts and the absence of cultural-linguistic barriers – underlining the importance of the earlier excerpt (sub-chapter 5.2.1), in which he notes the difficulties related to studying Greek as a foreign language.

Also noteworthy is LP's mention of return migration and secondary movement, with emigrants who gained economic success in Greece later moving to Russia and others returning to Georgia (both, presumably, due in large part to the 2008 economic crisis, as discussed in section 5.2). LP's assessment of these two phenomena in terms of success and failure, respectively, are particularly interesting in theoretical terms. While neoclassical theory views return migration as a sign of individual failure, whereby despite an improved economic situation the emigrant was unable to fully integrate into the destination country. The New Economics of labor migration, on the contrary, considers return migration an indicator of success, showing that the emigrant has achieved financial gain and then chooses to invest it in their home country. According to LP's story, however, despite their success, some emigrants still preferred to invest their gains not in Georgia, their country of origin, but in Russia. This can only be explained by world systems theory, with its attention to the ethnopolitical and sociocultural legacy of the Soviet Union, as discussed above.

GA, visiting from Cyprus, also spoke about the experience of alienation in a transformed Georgia, and raised the issue of local language barriers:

NL: In your opinion, what is a positive side to living in Georgia?
GA: Here, in Ts'alk'a, I don't see any plus sides. I can't even say I've been living in Georgia, I'm just passing through for now, so to speak. I can say that it's very beautiful here. There must be some advantages, of course...
NL: For example, you said earlier that, given the opportunity, you would live in Tbilisi. Why does Tbilisi appeal to you?
GA: You know what it is? I flew from Cyprus to Greece. From there I went to Istanbul by bus and from Istanbul, I came to Georgia. As soon as I set foot on Georgian soil, I noticed that I was finding it hard to communicate with people, mostly Georgians. For example, you ask them something and they answer in Georgian right away.
NL: You mean when you ask them in Russian?
GA: Yes, I don't know Georgian. Right now, I feel like a stranger in Ts'alk'a, because wherever I go, they answer me in Georgian. We can't find a common language. They don't understand me – I don't understand them. I'm like a foreigner here.

GA, male, 32-years-old, Ts'alk'a municipality.

The interviews with GA and LP make evident the importance of cultural, and not only economic, factors driving Greek Georgian emigration – particularly secondary movement from Greece to Russia – and limiting return migration. While linguistic and/or cultural alienation in post-Soviet Georgia was less important than unemployment or wages in driving initial migration, it can present critical obstacles for return migration. Hence, instead of being drawn back home, as New Economics theory would expect, successful Greek Georgian emigrants are drawn instead to Russia – a fact which only world systems theory can explain.

Chapter VI. Discussion

This chapter begins by dedicating closer attention to the history of Soviet repression and mass deportations as reflected in the analysis of the fieldwork data. It then revisits the book's theoretical synthesis in the context of the emigration, return migration and secondary movement vectors between Georgia, Greece and Russia. On this basis, it problematizes the tendency to amplify nationalism in interpretation of post-Soviet migration processes.

Soviet repression and deportations were not initially a focus of this study, neither for me nor for Concha Maria Höfler. During the fieldwork, however, Pontic interviewees in particular drew attention to this tragic episode, independently from the questionnaire. Accordingly, I decided to add a section to this subject which, it turned out, academics had yet to study in depth. The analysis of the interviews in this regard completes the picture, providing a wider perspective on Georgian Greeks' historical background and migration processes. For the present study, it was necessary to discuss Soviet deportations especially because they resulted in the 1939–59 censuses showing for the first time a decline in the Greek population in Georgia.

In analyzing the interviews, the main finding is a dissimilarity in the collective memory of Pontic and Urum Greeks, who have different levels of knowledge about these events. As shown in Chapter IV, the majority of Urum interviewees had limited knowledge, if any, about Soviet repression and deportations, whereas Pontic Greek respondents whose families were affected by them provided detailed accounts. Others had more fragmentary knowledge.

This difference can be explained by several factors. First, as noted above, the deportations were geographically limited to the coastal regions of Western Georgia, where Pontic Greeks lived in compact settlements. The repression not only substantially changed the ethnic composition of Adjara and Abkhazia, but also involved the shut-down of Greek-speaking schools, newspapers and cultural institutions. Second, the repressive reality of the Soviet system made this topic taboo until the second half of the 1980s, remembered only by the victims and their close relatives. As a result, information about these events did not circulate widely. Third, to this day the Georgian State does not officially acknowledge the victims of ethnic persecution in Soviet Georgia. In 1997 it introduced the "Law on the acknowledgement of citizens of Georgia as victims of political repression and social protection of repressed persons", which in its very first article excluded "persons who belong to the ethnic groups deported during the period from 25 February 1920 to 28 October 1990, and whose procedures for the rehabilitation shall be determined separately" (Legislative Herald of Georgia 2019).

Regarding post-Soviet mass emigration, the present study found no significant differences between Pontic and Urum Greeks. Quantitative and qualitative data analysis indicates that for both groups (as for the Georgian population as a whole), the decision to emigrate was unambiguously linked to the cataclysmic economic and political collapse of the 1990s. Joining previous studies on Georgian Greek migration (Voutira 2003; Diamanti-Karanou 2003; Sideri 2006; Kaurinkoski 2008, 2010), the distinguishing feature of the present work lies in analyzing this process using a theoretical synthesis of approaches to international labor migration, along with an in-depth analysis of the socio-economic transformations in Georgia which affected the migration processes.

The need for a synthesis of theoretical approaches was also identified through the fieldwork. While the motivations linked to labor migration are clearly indicated by the quantitative measures illustrated in Figure 10 and the interview excerpts presented in Chapter 5, each interviewee stressed different factors. Taken together, these factors could not all be accounted for by any of the individual theoretical models. Thus, EM (Section 5.1) is the only respondent whose description of her family member's emigration can be plainly linked to neoclassical theory. The rest of the interviewees mentioned motives accounted for by different theories in combination. For instance, AK's story (Section 5.1) divided the emigration into two main stages – temporary labor migration which is explained well by New Economics, and a cost-benefit calculation for the second stage, corresponding to the neoclassical model.

Other interviewees also drew attention to social capital and other factors which add significance to the insights of dual/segmented labor market theory (LP Sections: 5.1.1, 5.3.2 and 5.3.3; GN: 5.1.1 and 5.2.1; GA: 5.3.1 and 5.3.3; AL: 5.3.2). Therefore, while the fundamental factor behind Greeks' emigration from Georgia was and remains the difficult economic situation, the decision-making process was influenced by a variety of factors which are interrelated rather than mutually exclusive. At different points, the same interviewees described individual cost-benefit calculations, diversification of income and risk reduction for the household, as well as social capital and family ties.

In this regard, the development of emigration networks is central to the understanding of its driving factors (Massey 1999). The main emigration vectors are described in sub-chapter 5.2 and Maps 10–11). Here, an important finding is that both Russia and Greece were the main destinations, with Greece often serving as a way-station from which secondary migration took place – mainly to the EU, but also to Russia. This finding confirms the hypothesis that the mass emigration of Greeks from Georgia is in essence a case of labor migration. Co-ethnic factors only became significant insofar as they made Greek "repatriation" and thus European citizenship easily available, amid urgent economic conditions, along with the financial, educational, and legal support provided to Greeks from the post-Soviet space. However, from 2001, as Greece toughened its "repatriation" policy, the legal procedures related to emigration became significantly more complicated for Georgian Greeks. Nevertheless, emigration continued with the Georgian Greek population dropping from 15,000 in 2002 to 5,500 in the 2014 census. Therefore, motivations associated with labor migration continued to influence thousands of emigrants, even after ethnic identity ceased to confer any immediate advantage.

Map 9, and the analysis of interviews with GN (Section 5.2.1) and LP (Section 5.3.2), also work against the co-ethnic migration hypothesis. Firstly, there is very little return migration to Georgia, a fact substantiated by census data showing the steady and sizable decline of the Georgian Greek population over decades. Secondly, GN and LP speak about the advantages of a Greek – or rather European – citizenship primarily in terms of secondary movement. Both point to Greek-Georgian ethnic identity as a minor factor in the determination of migration patterns.

This leads us to the most interesting finding, namely the importance of the migration vector from Greece to Russia. While in the absence of official statistics the volume of such migration cannot be quantified, and out interviews do not constitute a representative sample, interviewees' stories were highly compelling in this respect. The explanation I would suggest for this finding is that emigration from Greece to Russia should, in fact, be considered return migration – albeit to the wider post-Soviet space rather than to Georgia specifically. This interpretation is supported by the fact that the main vectors of secondary movement from Greece are directed towards European countries, which Greek citizenship made legally accessible to Georgian Greek emigrants. None of the pull factors that influence this secondary movement apply to the Greece-Russia vector. Instead, LP speaks about "successful emigrants" who go to Russia after accumulating financial capital in Greece. Considering the cultural, social, and economic ties built during the Soviet period described in Section 5.3.3, the re-emigration successful Georgian Greeks from Greece to Russia actually corresponds to the New Economics account of return migration – but only if we take into account the wider socio-cultural legacies highlighted by world systems theory. This is strongly illustrated by the sense of alienation described by GA (Section 5.3.1), after the majority of Greeks had left Georgia and the Russian language had lost its dominant status there. Additionally, since the majority of Georgian Greeks are from rural areas, their experience of urban living in Greece and elsewhere may have made a return to a rural home less appealing. Consequently, we should widen the New Economics account of return migration to account for the fact that, at least for post-Soviet ethnic minorities, the concept of a "return destination" may be designated by socio-cultural factors that do not correspond to nominal political borders.

In conclusion, when analyzing the impact of socio-economic transformations on the migratory movements of Georgia's Greek community, it is essential to avoid methodological nationalism which uncritically assumes the primacy of co-ethnic factors. As the discussion has demonstrated, these have been secondary to economic factors in the present case study. This does not, however, mean adopting a transnational "boundlessness" (Wimmer and Schiller 2003) – since ethno-cultural and linguistic factors certainly play a supporting role, whether as drivers of state action (mass deportations, russification, Greek "repatriation" policy) or as factors in the choice of emigration and return destinations.

Chapter VII. Conclusion

The objective of this research was to study the mass migration processes of the Greek population of Georgia and to argue for the primary role of socio-economic rather than ethno-national factors in this process. It is also the first research project analyzing data on the Pontic and Urum Greek communities through a synthesis of labor migration theories. The use of both quantitative and qualitative methods to process the material from our fieldwork provided an optimal understanding of the circumstances that stimulate emigration, its routes, and the experiences of emigrants and their family members.

The research has established that the Greek vector of emigration should not be considered squarely within the context of the ethnic migrations that took place in the post-Soviet period. Rather, in the case of Georgian Greeks, the main driving factor was the all-encompassing economic and political collapse and insecurity that followed Georgia's independence. In this context, the ethnocultural factor only provided instrumental advantage due to Greek government policy, rather than being a primary motivator.

The decision to emigrate was made not only as an individual cost-benefit calculation, discussed by neoclassical theory, but also in a household context highlighted by the New Economics of labor migration and in the two structural contexts described by dual/segmented labor market theory and by world systems theory. The first informs our understanding of emigration to Greece, whose "repatriation" policy was in large part driven by the demand for cheap labor. This resulted in a dual/segmented labor market, with post-Soviet Greeks enjoying the relative "privilege" to join its lower strata. World systems theory, for its part, highlights the importance of the Russo-Georgian socio-economic and cultural context, which bears the attributes of a metropole-periphery relationship. This supports the significant insight that Russia, in this case study, constitutes a return destination for Georgian Greeks, opening the way for a broader theoretical understanding of return migration.

The study also established the space for further research on secondary and return migration from Greece, which has been identified but not examined in quantitative terms. A full-scale research project on the directions and causes of Greek Georgians' further migration from Greece is likely to lead to significant new findings. Meanwhile, we can certainly note the importance of the large emigration network revealed by the interview data. Chronologically, these connections constitute a second-phase development. In the earlier years of mass labor emigration, the decision was mainly taken on the individual and household levels. Pull factors significantly influenced decisions to migrate to Greece and Russia. Later on, as communication networks were established and became denser, they emerged as a pull factor in their own right. It is through such well-established network that emigrants' social capital became significant, reducing the expenses associated with emigration and facilitating emigrants' social mobility in the destination country.

Finally, as census data shows, the emigration of Greeks from Georgia is likely irreversible. Neither the census data nor the data gathered during fieldwork suggest any substantial return migration to Georgia, with Russia standing in for the return destination. Establishing oneself, after several years of separation, in a transformed Georgia proved quite difficult even for first-generation emigrant returnees. In the context of Georgia's current negative net migration rate and continuing economic difficulties, its Pontic and Urum Greek communities cannot be expected to regain their former place in the country's demographic and cultural landscape.

Bibliography

All-Union Population Census (1926a) Nacional'nyj sostav naselenija po regionam respublik SSSR (National composition of the population by regions of the republics of the USSR. Translation NL). URL: http://www.demoscope.ru/weekly/ssp/sng_nac_26.php?reg=2330

All-Union Population Census (1926b) Nacional'nyj sostav naselenija po regionam respublik SSSR (National composition of the population by regions of the republics of the USSR. Translation NL). URL: http://www.demoscope.ru/weekly/ssp/sng_nac_26.php?reg=2346

All-Union Population Census (1939a) Nacional'nyj sostav naselenija po respublikam SSSR (National composition of the population by the republics of the USSR. Translation NL). URL: http://www.demoscope.ru/weekly/ssp/sng_nac_39.php?reg=4

All-Union Population Census (1939b) Nacional'nyj sostav naselenija rajonov, gorodov i krupnyh sel sojuznyh respublik SSSR (National composition of the population of districts, cities and large villages of the union republics of the USSR. Translation NL). URL: http://www.demoscope.ru/weekly/ssp/ussr_nac_39_ra.php?reg=823

All-Union Population Census (1959a) Nacional'nyj sostav naselenija po respublikam SSSR (National composition of the population by the republics of the USSR. Translation NL). URL: http://www.demoscope.ru/weekly/ssp/sng_nac_59.php?reg=8

All-Union Population Census (1959b) Gorodskoe i sel'skoe naselenie oblastej respublik SSSR (krome RSFSR) po polu i nacional'nosti (Urban and rural population of regions of the republics of the USSR (except for the RSFSR) by gender and nationality. Translation NL). URL: http://www.demoscope.ru/weekly/ssp/resp_nac_59.php?reg=62

All-Union Population Census (1970a) Nacional'nyj sostav naselenija po respublikam SSSR (National composition of the population by the republics of the USSR. Translation NL). URL: http://www.demoscope.ru/weekly/ssp/sng_nac_70.php?reg=6

All-Union Population Census (1970b) Gorodskoe i sel'skoe naselenie oblastej respublik SSSR (krome RSFSR) po polu i nacional'nosti (Urban and rural population of regions of the republics of the USSR (except for the RSFSR) by gender and nationality. Translation NL). URL: http://www.demoscope.ru/weekly/ssp/resp_nac_70.php?reg=68

All-Union Population Census (1979a) Nacional'nyj sostav naselenija po respublikam SSSR (National composition of the population by the republics of the USSR. Translation NL). URL: http://www.demoscope.ru/weekly/ssp/sng_nac_79.php?reg=6

All-Union Population Census (1979b) Gorodskoe i sel'skoe naselenie oblastej respublik SSSR (krome RSFSR) po polu i nacional'nosti (Urban and rural population of regions of the republics of the USSR (except for the RSFSR) by gender and nationality. Translation NL). URL: http://www.demoscope.ru/weekly/ssp/resp_nac_79.php?reg=71

All-Union Population Census (1989a) Nacional'nyj sostav naselenija po respublikam SSSR (National composition of the population by the republics of the USSR. Translation NL). URL: http://www.demoscope.ru/weekly/ssp/sng_nac_89.php?reg=6

All-Union Population Census (1989b) Gorodskoe i sel'skoe naselenie oblastej respublik SSSR (krome RSFSR) po polu i nacional'nosti (Urban and rural population of regions of the republics of the USSR (except for the RSFSR) by gender and nationality. Translation NL). URL: http://www.demoscope.ru/weekly/ssp/resp_nac_89.php?reg=65

Arango, Joaquín (2000) Explaining migration: a critical view. *International social science journal*, volume 52 (165), 283–296.

Arango, Joaquín. (2004). Theories of international migration. In: Danièle Joly (ed), *International migration in the new millennium*, London: Routledge, 25–45.

Archvadze, Ioseb (2017) shromiti migrant'ebis mier tankhebis rep'at'riatsiis bunebisa da massht'abebis shesakheb (sakartvelosa da misi sakhmeleto mezoblebis magalitze) (The nature and scope of repatriation of funds by labor migrants *example of Georgia and neighboring countries). *Migratsia* 8: 79–95 (In Georgian)

Arel, Dominique (2003) Fixing ethnicity in identity documents: The rise and fall of passport nationalism in Russia. *Canadian Review of Studies in Nationalism* volume 30: 125–136.

Arel, Dominique (2006) Introduction: Theorizing the politics of cultural identities in Russia and Ukraine, in: Dominique Arel and Blair A. Ruble (eds.), *Rebounding Identities. The Politics of Identity in Russia and Ukraine*, Baltimore: Johns Hopkins University Press, 1–30.

Babalashvili, Ivane (1977) *Gruzinskaja SSR v gody Velikoj Otechestvennoj vojny 1941–1945 godov*. (Georgian SSR during the Great Patriotic War of 1941–1945. Translation NL). Tbilisi: Sabch'ota Sakartvelo.

Berggoltz, Olga (2011) *Zapretny Dnevnik [Forbidden Diary]*. Saint-Petersburg: Azbuka-Klassika Publ.

Bourdieu, Pierre & Wacquant, Loïc JD (1992) *An invitation to reflexive sociology.* Chicago: University of Chicago press.

Boyle, Paul; Halfacree, Keath & Robinson, Vaughn (1998) *Exploring Contemporary Migration.* London: Longman.

Brettell, Caroline B., & Hollifield, James F. (eds.) (2008) *Migration Theory. Talking Across Disciplines.* New York: Routledge.

Brubaker, Rogers (1998) Migration of ethnic unmixing in the 'New Europe'. *International Migration Review* volume 32 (4): 1047–1065.

Cassarino, Jean-Pierre (2004) Theorising Return Migration: The Conceptual Approach to Return Migration Revisited. *International Journal on Multicultural Societies,* 6(2): 253–279.

Chelidze, Natia (2006) *Shromiti emigratsia p'ost' sabch'ota saqartvelodan (Labor migration from post-Soviet Georgia).* Tbilisi: Center for Migration Studies, Ivane Javakhishvili Tbilisi State University (in Georgian).

Courtois, Stéphane; Werth, Nicolas; Panné, Jean-Louis; Paczkowski, Andrzej; Bartosek, Karel & Margolin, Jean Louis (1999) *The Black Book of Communism: Crimes, Terror, Repression.* Cambridge: Harvard University Press.

CRRC (2007) Migration and Return in Georgia: Trends, Assessments, and Potential. *Report submitted for the Danish Refugee Council by Caucasus Research Resource Centres (CRRC)* Georgia. 7–120.

Danilov, Viktor (2011) Stalinizm i sovetskoe obshchestvo (Stalinism and Soviet society]. *Istoriya krestyanstva Rossii v XX vek: Izbrannye trudy. [The History of Russian Peasantry in the 20th Century. Selected Works]* Vol. 2, Moscow: ROSSPEN, pp. 710–723 (in Russian).

De Haas, Hein, Mark J. Miller, and Stephen Castles (2019) *The age of migration: International population movements in the modern world.* Red Globe Press.

De Haas, Hein; Fokkema Tineke & Fassi Fihri, Mohamed (2015) Return migration as failure or success?. *Journal of international migration and integration,* volume 16(2): 415–429.

Dershem, Larry & Khoperia Tea (2004) *The Status of Households in Georgia—2004.* Tbilisi: Save the Children, USAID, and the Institute for Polling and Marketing.

Diamanti-Karanou, Panagoula (2003) Migration of Ethnic Greeks from the Former Soviet Union to Greece, 1990–2000: Policy Decisions and Implications. *Journal of Southeast European and Black Sea Studies,* volume 3 (1): 25–45.

Dzhukha, Ivan (2006) *Grečeskaja operacija. Istorija repressij protiv grekov v SSSR (The Greek operation: The History of repressions against Greeks in the USSR).* St. Petersburg: Altheia (in Russian).

EI-LAT (2012) *saqartvelos ek'onomik'uri transformatsia: damouk'ideblobis 20 ts'eli (Georgia's Economic Transformation: 20 years of independence).* Tbilisi: Interim report of European Initiative Liberal Academy. (in Georgian)

Ellman Michael (2007) Stalin and the Soviet Famine of 1932–33 Revisited. *Europe-Asia Studies*, volume 59(4): 663–693.

Fanon, Frantz (1961) *The Wretched of the EarthTrans. Richard Philcox.* New York: Grove (2004).

Federal State Statistics Service of Russia (2019) *Oficial'naja statistika: Rynok truda, zanjatost' i zarabotnaja plata (Official statistics: labor market, employment, wages. Translation NL)* last accessed 02.05.2019 URL: http://www.gks.ru/wps/wcm/connect/rosstat_main/rosstat/ru/statistics/wages/

Forni, Floreal H.; Freytes Frey, Ada & Quaranta, Germán (1998) Frédéric Le Play: a forefather of social economics. *International Journal of Social Economics* volume 25(9): 1380–1397.

Gachechiladze, Revaz (1997) *mosakhleobis migratsia saqartveloshi da misi sotsialur-ek'onomik'uri shedegebi (Population migration in georgia and social-economical consequences).* Tbilisi: United Nations development program (in Georgian).

Gachechiladze, Revaz (2008) *Chemi meotse sauk'une (My XX century) vol I-II. Ed. Qitoshvili, Tea.* Tbilisi: Bakur Sulakauri Publishing (in Georgian).

Garakanidze, Irine (2000) Sakartveloshi berdznebis gadmosakhlebis ist'oria (XVIII-XXcc.) (History of Greek Resettlement in Georgia (XVIII-XXcc). Translation NL). In: Gordeziani, Rismag (ed.) *Berdznebi saqartveloshi* (Greeks in Georgia. Translation NL). 2nd edition. Tbilisi: Logosi, 28–71 (in Georgian).

GEOSTAT (1999) *sakartvelos st'at'ist'ik'is erovnuli sasakhuris 1999 ts'lis saqmianobis angarishi* (Annual Report of the National Statistics Office of Georgia for 1999. Translation NL). Tbilisi: Geostat

GEOSTAT (2003) sakartvelos mosakhleobis 2002 ts'lis p'irveli erovnuli sak'oveltao aghts'eris shedegebi (Results of the first national census of Georgia in 2002. Translation NL). Volume 1, Tbilisi: Geostat

GEOSTAT (2014) *Statistical Yearbook of Georgia.* Tbilisi: National Statistics Office of Georgia.

GEOSTAT (2016) *Statistical Yearbook of Georgia.* Tbilisi: National Statistics Office of Georgia.

GEOSTAT (2019) *Statistical Yearbook of Georgia.* Tbilisi: National Statistics Office of Georgia.

GEOSTAT (2020) *Indicators of Employment and Unemployment Calculated Based on New Standards of the International Labour Organization.* URL:

https://www.geostat.ge/en/single-news/2009/indicators-of-economic-activity-of-the-population-employment-and-unemployment

Giorgadze, Marine (ed); Mgeladze, Nugzar; Tunadze, Temur; Dzenaladze, Nino; Loria, Manuchar; Abuladze, Teona & Chkhartishvili-Masmanidi, Sophiko (2017) *Pontoeli berdznebi ach'arashi: ts'arsuli da tanamedroveoba (Pontic Greeks in Ajara: past and present)*. Batumi: Batumi Shota Rustaveli State University, Faculty of Humanities (in Georgian).

Glaser, Barney G. & Strauss, Anselm L. (1967) *Discovery of grounded theory: Strategies for qualitative research*. New Brunswick and London: AldineTransaction.

Grenoble, Lenore A. (2003) *Language Policy in the Soviet Union*. Dordrecht: Springer Netherlands

Gugushvili, Alex (2011) Understanding Poverty in Georgia. *Caucasus Analytical Digest* volume 34: 15–18.

Hellenic Ministry of Foreign Affairs (2019) *Visas for Greeks travelling abroad* last accessed 30.04.2019. URL https://www.mfa.gr/en/visas/visas-for-greeks-traveling-abroad/visas-for-greeks-travelling-abroad.html.

Hionidou, Violetta. (2012) "Abroad I was Greek and in Greece I am a foreigner": Pontic Greeks from former Soviet Union in Greece. *Journal of Modern Greek Studies*, volume 30: 103–127.

Höfler, Concha Maria (2020) *Boundaries and Belonging in the Greek Community of Georgia*. Nomos Verlagsgesellschaft mbH & Co. KG

Höfler, Concha Maria, Stefanie Böhm, Konstanze Jungbluth, and Stavros Skopeteas (2016) "Urum and Pontic Greeks: Communities and language situations." *STUF-Language Typology and Universals*, volume 69(2): 171–174.

Hosking, Geoffrey A. (1997) *Russia: people and empire, 1552–1917*, Cambridge: Harvard University Press.

Human Development Reports (2019) *Human Development Data 1990–2017*, last accessed 30.04.2019. URL http://hdr.undp.org/en/data.

Iashvili, Ia - (2018) sazogadoebrivi geografia da migratsia (Human geography and migration. Translation NL)., In: Violeta Wagner (ed.) *Migratsiis Sakhelmdzghvanelo* (Migration Textbook). Tbilisi: International Centre for Migration Policy Development, 190–255 (in Georgian).

Iashvili, Ia; Gvenetadze, Tinatin; Mikautadze, Sali; Davituliani, Tsitsino & Dogonadze, Tatia (2014) Split families and family members left behind: Migration from Georgia. *Humanities and Social Sciences Review* 3(2): 345–356.

IMF (International Monetary Fund) (2003) *Economic Development and Poverty Reduction Programme of Georgia*. Tbilisi: s.n. (in Georgian).

Ivakhnyuk, Irina (2007) Eurasian migration system: theoretical and political approaches. *International Migration of Population: Russia and the Contemporary World.* Vol 20: 81–100.

Jalabadze, Natia (2011) Some aspects of intercultural communication in Kvemo Kartli. *Klio*, volume 7: 559–579. (in Georgian).

Janiashvili, Lavrenti (2008) etno-k'ulturuli protsesebi sakartveloshi (Ethnocultural processes in Georgia. Translation NL) In: Basilaia, Merab (ed.) *Etnosebi saqartveloshi* (Ethnics in Georgia). Tbilisi: Tolerance center under the Auspices of the Public Defender, 7–23 (in Georgian).

Jaoshvili, Vakhtang (1984) *sakartvelos mosakhleoba XVIII-XIX saukuneebshi* (Population of Georgia in XVIII-XIX centuries. Translation NL). Tbilisi: Metsniereba (in Georgian).

Jaoshvili, Vakhtang (1996) *saqartvelos mosakhleoba: ist'oriul-demografiuli da ek'onomik'ur geografiuli gamok'vleva (Population of Georgia: Historic-demographyc and economic-geographic research. Translation NL).* Tbilisi: Metsinereba (in Georgian).

Jishkariani, David (2018) gadasakhleba davits'k'ebis udabnoshi (Exile – the desert of oblivion). German Georgian Archive, last accessed 23.02.2021. URL http://german-georgian.archive.ge/ka/blog/59

Junge, Marc & Müller, Daniel (2017) Nation Building by Terror in Soviet Georgia 1937–1938. In: Andrej Kotljarchuk & and Olle Sundström (eds.), *Ethnic and Religious Minorities in Stalin's Soviet Union.* Huddinge: Södertörns academic studies, 123–152.

Junge, Marc; Bonwetsch, Bernd & Müller Daniel (2015) *etnosi da t'erori sakartveloshi* (*Ethnos and Terror in Soviet Georgia).* Tbilisi: Intelekti (in Georgian).

Junge, Marc; Tushishvili, Omar & Bonwetsch, Bernd (2015) *bolshevikuri ts'esrigi sakartveloshi, Tomi I, didi t'erori k'avk'asiis p'at'ara resp'ublik'ashi* (Bolshevik Order in Georgia. Vol. I, The Great Terror in a small Caucasian republic. Translation NL). Tbilisi: s.n. (in Georgian)

Kaukhchishvili, Simon (1942) Berdznebis dasakhlebis ist'oria saqartveloshi (The history of Greek resettlement in Georgia. Translation NL). *kutaisis p'edagogiuri inst'it'ut'is shromebi,* Volume 4: 219–239 (in Georgian).

Kaurinkoski, Kira (2008) Migration from Ukraine to Greece since perestroika: Ukrainians and 'returning' ethnic Greeks. Reflections on the migration process and on collective identities. *Migrance,* Special issue on Greece, volume 31: 71–85.

Kaurinkoski, Kira. (2010). Privileged co-ethnic Greek migrants from the former Soviet Union in the greater Athens area: reflections on individual and collective integration strategies into Greek society. In: Jasna Capo-Zmegac;

Christian Voss & Klaus Roth (eds.), *Co-Ethnic Migrations Compared. Central and Eastern European Contexts,* Berlin: Peter Lang.

Khorava, Bezhan (2004) *afkhazta 1867 ts'lis muhajiroba* (Abkhaz Muhajirism in 1867. Translation NL) Tbilisi: Artanuji.

Khutsishvili, Lia & Papuashvili, Nugzar (2008) dukhoborebi (Dukhobors. Translation NL) In: Kiknadze, Zurab (ed.) *religiebi saqartveloshi* (Religions in Georgia). Tbilisi: Tolerance center under the Auspices of the Public Defender, 231–238 (in Georgian).

King, Russel; Lazaridis, Gabriella & Tsardanidis, Charalambos (eds.) (2000) *Eldorado or Fortress? Migration in Southern Europe*, London: Macmillan.

King, Russell (2012) *Theories and typologies of migration: an overview and a primer.–Willy Brandt Series of Working Papers in International Migration and Ethnic Relations 3/12*, Malmö: Malmö Institute for Studies of Migration, Diversity and Welfare (MIM).

Kitromilidis, Paschalis, M. (1990) Greek irredentism in Asia Minor. *Middle Eastern Studies* 26(1). 3–17.

Kokoev, Konstantin; Svanidze, Guram & Melikishvili, Liana (1999) *National Minorities in Georgia.* Tbilisi: Intelecti, centre for the Protection of Constitutional Rights in agreement with UNHCR.

Komakhia, Mamuka (2011) *etnik'uri ruk'is formireba saqartveloshi (formation of an ethnic map in Georgia),* Tbilisi: BTTK Policy Research Group. (in Georgian)

Kurekova, Lucia. (2011) *Theories of migration: Conceptual review and empirical testing in the context of the EU East-West flows. Paper prepared for Interdisciplinary conference on Migration. Economic Change, Social Challenge.* April 6–9, 2011, University College London.

Labbate, Gabriel (1999) *Human Development Report, Georgia.* Tbilisi: Nekeri.

Latsabidze, N., Tsartsidze M. (2013) Saqartvelos soflis mosaxleobis efeqt'iani dasaqmebis uzrunvelqofis gzebi (Effective Employment of Georgia's Rural Population). Bioeconomy and Sustanable Development of Agriculture. II International Scientific-Practical Conference. Proceedings. Tbilisi: TSU Publishing House, 271–276 (in Georgian).

Layton, Susan (2005) *Russian literature and empire: Conquest of the Caucasus from Pushkin to Tolstoy*, Cambridge: Cambridge University Press.

Lee, Everett S. (1966). A Theory of Migration. *Demography*, volume 3(1): 47–57.

Legislative herald of Georgia (2019) Law of Georgia on the acknowledgement of citizens of georgia as victims of political repression and social protection of repressed persons, last accessed 10.05.2019. URL https://matsne.gov.ge/en/document/view/31408?publication=11

Lewis, W. Arthur (1954) Economic Development with Unlimited Supplies of Labour. *The Manchester School*, 22(2): 139–191

Loladze, Nika (2016) Greeks of Georgia: Main factors and motivations of emigration. *STUF-Language Typology and Universals*, *69* (2): 175–192.

Lucas, Robert E. & Stark, Oded (1985). Motivations to remit: Evidence from Botswana. *The Journal of Political Economy*, volume 93(5): 901–918.

MacDonald, John S. & MacDonald, Leatrice D. (1964). Chain Migration, Ethnic Neighbourhood Formation and Social Networks, *Millbank Memorial Fund Quarterly*, volume 42(1): 82–97.

Mace, James E. (1986) The Man-Made Famine of 1933 in Soviet Ukraine. In: Roman Serbyn & Bohdan Krawchenko (eds.), *Famine in Ukraine 1932–1933*, Edmonton: Canadian Institute of Ukrainian Studies, University of Alberta, 1–14.

Maddison, Angus (2003) *The World Economy: Historical Statistics*, Paris: OECD.

Massey, Douglas S. (1999) "Why does immigration occur? A theoretical synthesis." In: Charles Hirschman; Philip Kasinitz & Josh DeWind (eds.), *The Handbook of International Migration*, London: The Russell Sage Foundation, 43–52.

Massey, Douglas S.; Arango, Joaquin, Hugo, Graeme; Kouaouci, Ali & Pellegrino, Adela (1999) *Worlds in Motion: Understanding International Migration at the End of the Millennium*. Oxford: Clarendon Press.

Massey, Douglas S; Arango, Joaquin; Hugo, Graeme; Kouaouci, Ali; Pellegrino, Adela & Taylor, J. Edward (1993) Theories of International Migration: A review and Appraisal. *Population and Development Review*, volume 19(3): 431–466.

Mavrogordato, John (1931) *Modern Greece: A Chronicle and a Survey, 1800–1931*. London: Macmillan.

Melikishvili, Lia & Jalabadze, Natia (2016) The issue of ethnic identity and aspects of cross-cultural orientation of the Greeks in Georgia (the example of Ts'alk'a Greeks). *STUF - Language Typology and Universals*, volume 69: 193–212.

Melikishvili, Lia (2011) *Sotsialuri usafrtkhoebis etnik'uri asp'eqt'ebi polieTnik'ur sazogadoebashi* (Ethnic aspects of social sequrity in polyethnic society) Tbilisi: Mtsignobari (in Georgian).

Mikiashvili, Lela (1989) *Oman Kherkheulidze – "mefoba irakli meorisa"* (Oman kherkheulidze – "The reign of the king Heraclius II". Translation NL); Tbilisi: metsniereba.

Morawska, Eva (2007) *International Migration: Its Various Mechanisms and Different Theories that Try to Explain it*, Malmö: Malmö University, Willy Brandt Series of Working Papers in International Migration and Ethnic Relations.

Natmeladze, Makvala (2002) *Migration Processes in Georgia in the 1940's (The expulsion of Turks and other demographic groups from Georgia from 1941 to 1951).* Tbilisi: CIPDD (in Georgian).

NDI (National Democratic Institute) (2015) *Public Attitudes in Georgia: Results of a April 2015 survey carried out for NDI by CRRC Georgia.* Last accessed 04.04.2019. URL https://www.ndi.org/April-2015-Public-Opinion-Poll-Georgia .

OECD (Organisation for Economic Co-operation and Development) (2019) *Average annual wages.* Last accessed 02.05.2019. URL https://stats.oecd.org/Index.aspx?DataSetCode=AV_AN_WAGE#.

Papuashvili, Nugzar (2008) malak'nebi (Molokans. Translation NL) In: Kiknadze, Zurab (ed.) *religiebi saqartveloshi* (Religions in Georgia). Tbilisi: Tolerance center under the Auspices of the Public Defender, 239–245 (in Georgian).

Pashaeva, Lamara (1983) *Nekotorye kalendarnye prazdniki Urumov* (Some Calendar Holidays of Urums. Translation NL) Tbilisi: Matsne (in Russian).

Pashaeva, Lamara (1992) *Sem'ja i semejnyj byt grekov Tsalkinskogo raiona* (Family and a family life of Tsalka region's Greeks. Translation NL). Tbilisi: Metsniereba (in Russian).

Piore, Michael J. (1979) *Birds of passage: Migrant labor and industrial societies.* New York: Cambridge University Press.

Pkhakadze, Manana & Berdzenishvili, Levan (2000) berdzenta q'ofa (Everyday life of Greeks. Translation NL). In: Gordeziani, Rismag (ed.) *Berdznebi saqartveloshi* (Greeks in Georgia. Translation NL). 2nd edition. Tbilisi: Logosi, 71–84 (in Georgian).

Pohl, J. Otto (2000) Stalin's genocide against the "Repressed Peoples". *journal of genocide research,* volume 2(2): 267–293.

Polian, Pavel (2003) *Against Their Will: The History and Geography of Forced Migrations in the USSR.* Budapest: Central European University Press.

Popov, Anton (2016) *Culture, Ethnicity and Migration After Communism: The Pontic Greeks,* London: Routledge.

Portes, Alejandro (2001) Introduction: the debates and significance of immigrant transnationalism. *Global Networks,* volume 1(3): 181–194.

Portes, Alejandro (2001) Introduction: the debates and significance of immigrant transnationalism. *Global Networks,* volume 1(3): 181–194.

Pratsinakis, Manolis (2014) Resistance and Compliance in Immigrant–Native Figurations: Albanian and Soviet Greek Immigrants and their Interaction with Greek society. *Journal of Ethnic and Migration Studies, volume* 40(8): 1295–1313.

Ranis, Gustav & Fei, John C. (1961) A Theory of Economic Development. *American Economic Review*, volume 51(4): 533–565.

Ravenstein, Ernst G. (1885) The Laws of Migration – I. *Journal of the Statistical Society*, volume 48(2): 167–227.

Ravenstein, Ernst G. (1889) The Laws of Migration – II. *Journal of the Statistical Society*, volume 52(2): 214–301.

Rummel, Rudolph J. (2017) *Lethal politics: Soviet genocide and mass murder since 1917*. London: Routledge.

Said, Edward (1993) *Culture and Imperialism*. New York: Alfred A. Knopf.

Salukvadze Joseph and Meladze Giorgi (2014) Georgia: Migration, a Main Risk Towards Sustainable Demographic Future, for the publication supported by the International Visegrad Fund in frame of EASTmig project. *Discovering migration between Visegrad countries and Eastern Partners*: 150–169.

Savin, Andrey (2017) Ethnification of Stalinism? Ethnic Cleansings and the NKVD Order № 00447 in a Comparative Perspective. In: Andrej Kotljarchuk & Olle Sundström (eds.), *Ethnic and Religious Minorities in Stalin's Soviet Union*, Huddinge: Södertörns University: 47–67.

Shelia, Mzia (2016) shromiti emigratsiis zegavlena p'ost sabch'ota saqartvelos demograpiul da ek'onomik'ur ganvitarebaze (The Impact of Labor Emigration on the Demographic and Economic Development of Georgia in the Post-Soviet Period). *Migration*, volume 7: 189–201. (in Georgian).

Shelia, Mzia (2017) qalta shromiti emigratsia saqartvelodan (Labor migration of women from Georgia). *Migration* 8: 114–133. (in Georgian)

Sideri, Eleni (2007) Crossing b-orders: From Georgia to Greece. Female migration and female transformations, in: Yota Papageorgiou (ed.), *Gendering Transformations*, Heraklion: e-media University of Crete, 98–108.

Sideri, Eleni (2012) Looking for the 'language' of recognition among Greek communities of Georgia. *Anthropological Journal of European Cultures*, volume 21(1): 41–59.

Sideri, Eleni. (2006) *The Greeks of the former Soviet Republic of Georgia. Memories and practices of diaspora*. London: School of Oriental & African Studies, University of London.

Sjaastad, L.A. (1962). The Costs and Returns of Human Migration. *Journal of Political Economy*, volume 70(1): 80–93.

Slezkine, Yuri (1994) The USSR as a communal apartment, or how a socialist state promoted ethnic particularism. *Slavic Review*, volume 53: 414–452.

Solchanyk, Roman (1982) Russian language and Soviet politics. *Soviet Studies*, volume 34(1): 23–42.

Stark, Oded & Bloom, David E. (1985). The new economics of labor migration. *The American Economic review*, volume 75(2), 173–178

Stark, Oded (1991) *The migration of labor.* Oxford: Blackwell.

Sulaberidze, Avtandil (1999) Toward Poverty Eradication in Georgia. In: Yogesh Atal (ed.) *Poverty in Transition and Transition in Poverty: Recent Developments in Hungary, Bulgaria, Romania, Georgia, Russia, Mongolia.* New York: Berghahn Books, 130–176.

Sulaberidze, Avtandil (2017) migratsiis statistikuri agh'ritskhvis metodologiuri problemebi da misi shedegebi sakartveloshi (Migration statistics: methodological problems of accounting and its results in georgia). *Migratsia,* volume 8: 70–78

Suny, Ronald Grigor & Martin, Terry (Eds.) (2001). *A State of Nations: Empire and Nation-making in the Age of Lenin and Stalin.* Oxford: Oxford University Press.

Tellis, Winston M. (1997) Introduction to case study. *The qualitative report,* volume 3(2): 1–14.

Thomas, William I. & Znaniecki, Florian (1918–1920) *The Polish Peasant in Europe and America*, Boston: William Badger.

Todaro, Michael P. & Maruszko, Lydia (1987) Illegal migration and US immigration reform: A conceptual framework. *Population and development review,* volume 13(1): 101–114.

Topchishvili, Roland (2015) *Khalkhta migratsiebi (People's Migrations)*, Tbilisi: "Universali" publishing house. (in Georgian)

Troynitsky, Nikolay (1905) *Pervaja vseobŝaja perepis' naselenija Rossijskoj Imperii 1897g. (The first universal census of the population of the Russian Empire in 1897. Translation NL)* Vol. LXVI. Saint Petersburg: Ripol (In Russian).

Tsuladze, Giorgi; Sulaberidze, Avtandil (2015) *demografiis safudzvlebi* (Basics of Demography. Translation NL) Tbilisi: Ilia State University Press.

Tvalchrelidze, Alexander; Silagadze, Avtandil; Keshelashvili, Giuli & Gegia, D. (2011) *Georgia's Social Economic Development Program.* Tbilisi: Nekeri (in Georgian).

UGSRA (Department of State Statistics of the Republic of Abkhazia) (2019) *Abhazija v Cifrah za 2014 god* (Abkhazia in numbers for 2014. Translation NL) last accessed: 03.05.2019 URL: https://ugsra.org/abkhaziya-v-tsifrakh/2014-god.php (In Russian).

Voell, Stéphane (ed.) (2016) *Traditional Law in the Caucasus: Local Legal Practices in the Georgian Lowlands.* Marburg: Curupira

Voell, Stéphane; Jalabadze, Natia; Janiashvili, Lavrenti & Kamm, Elke (2014) Identity and traditional law: Local legal conceptions in Svan villages, Georgia." *Anthropological Journal of European Cultures,* volume 23(2): 98–118.

Voutira, Eftihia (2004) Ethnic Greeks from the former Soviet Union as 'privileged return migrants'. *Espace, populations, sociétés,* volume 3: 533–544.

Wallerstein, Immanuel (1974) *The Modern World-System: Capitalist Agriculture and the Origins of the European World-Economy in the Sixteenth Century*. New York: Academic Press.

Wallerstein, Immanuel (1979) *The Capitalist World Economy*. Cambridge: Cambridge University Press.

Wang, Jian-Ye (1999) *The Georgian hyperinflation and stabilization*. Washington DC: International Monetary Fund.

White, Paul & Woods, Robert (eds.) (1980) *The Geographical Impact of Migration*. London: Longman.

Wimmer, Andreas, & Glick Schiller, Nina (2003) "Methodological Nationalism, the Social Sciences, and the Study of Migration: An Essay in Historical Epistemology 1., *International migration review,* volume 37(3): 576–610.

World Bank (2021) *GDP per capita (constant 2010 US$) - Georgia*. World Bank national accounts data, and OECD National Accounts data files. Accessed 09.03.2021. URL: https://data.worldbank.org/indicator/NY.GDP.PCAP.KD?locations=GE

World Bank (2020) *Unemployment, total (% of total labor force) (modeled ILO estimate) - Russian Federation, Greece*. Accessed 09.03.2021. URL: https://data.worldbank.org/indicator/SL.UEM.TOTL.ZS?locations=RU-GR

Zelinsky, Wilbur (1971) The hypothesis of the mobility transition. *Geographical review,* volume 61(2): 219–249.

Zurabishvili, Tamar (2007) *Shromiti emigratsia daba tianetidan: emigraciuli qselebis ganvitareba (Labor migration from town of Tianeti: development of migration networks)*. Tbilisi: Ilia State University, Department of philosophy and social sciences. (in Georgian).

Zurabishvili, Tinatin (2006) *tvisebrivi metodebi sotsialur k'vlevashi* (Qualitative Methods in Social Research). Tbilisi: Center for Social Sciences (in Georgian).

Appendix

Appendix 1. Urum (*) and Pontic (•) Greek compact settlements in Georgia by municipalities

Kvemo Kartli						
Ts'alk'a Municipality			***Tetrits'q'aro mun.***	***Dmanisi mun.***	***Marneuli mun.***	***Bolnisi mun.***
*Akhalsheni (K'iriak'i)**	*•Gumbati*	*Sameba (Guniak'ala)**	*Abeliani**	*Dmanisi**	*Marneuli**	*Bolnisi**
*Arts'ivani (Jinisi)**	*Imera**	*•Santa*	*•Didi Iraga*	*Ganakhleba**	*Opreti**	
*Avranlo**	*K'okhta (Akhalik'i)**	*Saq'drioni (Edikilisa)**	*•Ivanovk'a*	*•Gora*	*Shulaveri**	
*Bareti (Bashkoi)**	*Khando**	*•T'arsoni*	*•Jigrasheni*	*Iailo**		
*Bediani**	*•Khareba (Neon-Kharaba)*	*T'beti (Khadik'i)**	*•P'atara Iraga*	*•Sak'ire*		
*Berta (Olianki)**	*Khramhesi**	*•Trialeti**	*Tetrits'q'aro**	*Sark'ineti**		
*Beshtasheni**	*Livadi**	*Ts'alk'a**	*Ts'ints'qaro**	*Velisp'iri**		
*•Chrdilisubani (Shepiaki)**	*Rekha**	*Ts'ints'q'aro**	*•Vizirovk'a*			
*Gantiadi (Tiakkilisa)**	*Sabech'isi (K'arak'omi)**	*Sameba (Guniak'ala)**	*Abeliani**			

Samtskhe-Javakheti		Abkhazia			Adjara		
Borjomi mun.	***Akhaltsikhe mun.***	***Sokhumi mun.***	***Gudauta mun.***	***Gulripshi mun.***	***Batumi mun.***	***Kobuleti mun.***	***Khelvachauri mun.***
•*Andezit'i*	•*Mikelts'minda*	•*Odishi*	•*Gudauta*	•*Gulripshi*	•*Batumi*	•*Ach'q'va*	•*Akhalsheni*
•*Tsikhisjvari*		•*Olginsk'oe*				•*Dagva*	
		•*Okt'omberi (Pavlovk'a)*				•*K'virik'e*	
		•*Sokhumi*				•*Kobuleti*	

Sprachliche Konstruktion sozialer Grenzen: Identitäten und Zugehörigkeiten / Linguistic Construction of Social Boundaries: Identities and Belonging

Herausgegeben von / Edited by Konstanze Jungbluth und Mônica Maria G. Savedra

Band 1 Dagna Zinkhahn Rhobodes: Sprechen entlang der Oder. Der Charakter der sprachlichen Grenzen am Beispiel der deutsch-polnischen Sprachroutine. 2016.

Band 2 Ekaterina Popova: Sprachen und interethnische Beziehungen in Estland in der Umbruchszeit. Russische Bevölkerungsgruppe zwischen Anpassung und Protest. 2016.

Band 3 Mario Luis Monachesi Gaio: Etnicidade Linguística em Movimento. Os Processos de Transculturalidade Revelados nos Brasileirítalos do Eixo Rio de Janeiro-Juiz de Fora. 2018.

Band 4 Rita Tamara Vallentin: Language and Belonging. Local Categories and Practices in a Guatemalan Highland Community. 2019.

Band 5 Débora Amaral da Costa: Identities of Migration. A Narrative-based Approach to the Studies of Social Representation. 2020.

Band 6 Layla Cristina Jochmann: Português como Língua de Herança. A coesão em textos escritos por crianças plurilíngues. 2020.

Band 7 Jan Aguiar Pöhlmann: Mehrsprachigkeit per Gesetz: Die Implementierung der Ko-Offizialisierung von Sprachen in Brasilien. 2021.

Band 8 Nika Loladze: Migratory Movements of Georgia's Greek Community. The Impact of Current Socio-economic Transformations. 2021.

www.peterlang.com

www.ingramcontent.com/pod-product-compliance
Ingram Content Group UK Ltd.
Pitfield, Milton Keynes, MK11 3LW, UK
UKHW041827031026
15587UKWH00002B/614

9 783631 847268